MUNICIPAL WASTE MANAGEMENT IN EUROPE

Environment & Management

VOLUME 10

The titles published in this series are listed at the end of this volume.

Municipal Waste Management in Europe

A Comparative Study in Building Regimes

Edited by

NICOLAS BUCLET

Centre International de Recherche sur l'Environnement et le Développement, Nogent sur Marne, France

and

OLIVIER GODARD

Laboratoire d'Econométrie, Ecole Polytechnique, Paris, France

Springer-Science+Business Media, B.V.

Library of Congress Cataloging-in-Publication Data is available.

ISBN 978-90-481-5292-6 ISBN 978-94-015-9476-9 (eBook)
DOI 10.1007/978-94-015-9476-9

Printed on acid-free paper

Originally published by Kluwer Academic Publishers in 2000.
Softcover reprint of the hardcover 1st edition 2000

Table of Contents

Introduction

Nicolas Buclet and Olivier Godard

In terms of economic scale, waste management is one of the two most important environmentally oriented sectors.[1] It stands at the cross-roads in the material organization of society, resource management, changing lifestyles and consumption patterns, and ecological issues. For many years waste management has been perceived as a resources and health issue, confined mainly to dense urban areas, and not an environmental issue. In contemporary affluent societies, however, the scale reached by waste flows, the inheritance of accumulated deposits in soils from the waste of previous generations and increasing levels of public concern about environmental protection and quality of life have all conspired to impose a fresh look at what waste really implies for a modern society. We are obliged to focus our attention on such questions as how the circulation of matter is at present organized by society and can be modified and controlled if economic development is to become more environmentally sustainable. This is the period we live in. Significant changes in waste management in European countries have been introduced during the last decade or so. To some extent the transition between traditional regimes mainly based on local disposal and new regimes based on a revised organisation of flows of waste matter is still in the making, involving new attitudes, new activities, new technologies and new incentives, reducing the pressure on virgin natural resources and eliminating the huge dissipation of various pollutants into the environment.

This book is devoted to municipal waste management. It embraces all waste flows emanating from the consumption of goods by inhabitants, shops, local industries, municipal services and so forth, all of which were historically managed within the context of local authorities. It does not consider every aspect of waste management and, specifically, it does not include industrial waste, whether ordinary or toxic. This self-imposed limitation lies in the fact that managing industrial waste raises specific problems. Above all, this kind of waste is mainly the province of the same, rather homogeneous, industrial culture, linking companies driven by a shared industrial rationale on the basis of engineering oriented expertise, although the resulting practices have increasingly been contested environmentally. In contrast, at the heart of municipal waste management lies the meeting, or collision, of diverse cultural outlooks and contrasting social rationales, involving heterogeneous actors: industrial companies, retailers and services, national administrations, local authorities,

N. Buclet and O. Godard (eds.), Municipal Waste Management in Europe, 1–4.

consumers and citizens (who are also both co-producers of waste and victims of nuisances), environmental NGOs and other forms of organizations involving concerned citizens. How are all these worlds interacting and meeting to address a common problem? How are the local realities of waste management, beyond a considerable autonomy in decision-making, framed by national institutional parameters and policies?

We will now consider the main theoretical justification for the comparative study of municipal waste management in five European countries: France, Germany, Greece, Italy and the Netherlands. When confronted with broadly similar problems, countries which also share similar economic organization and general cultural backgrounds have chosen rather different waste management regimes.[2] These differences extend to the institutional framework selected, the technology used and the processes through which the regimes have been developed. How can we make sense of these differences? From a rather naive economic viewpoint, if some technique or institutional mechanism is proven to be the best choice, we would expect to find it selected in the same way in each country. Why is this not the case? Good reasons immediately emerge: geography (climate, population density) and the level of economic development are two obvious explanations for differing choices. Obvious, but not sufficient. Beyond standard economic assumptions that choices can solely be explained by the selection of the best cost–benefit ratios, the study of waste management reveals that social choice in such a field depends on the way in which political processes and institutional mechanisms combine the heterogeneous rationales that co-exist in western societies. At the core of regimes lie basic conventions that simultaneously order social representations and qualifications, and the actions taken within a regime. These conventions are necessary to achieve co-ordination and give direction to a process of change; they are ultimately expressing some basic normative 'ideal' views about the organization of waste management that will prevail in the future; what we call, in the final chapter, 'myths' of waste management in a sustainable society.

Describing change in regimes, institutional parameters and policies raises the question of how change is being introduced. Can we find examples of radically new systems and instruments? Can we, instead, identify factors ensuring continuity through time? What is the driving force for change: a step by step incremental problem-solving strategy or more sweeping societal changes that lead to a re-appraisal of existing policies and programmes? Are the changes that have been introduced in different European countries leading them to converge progressively towards a common regime, or are they instead increasing differences between them, since the traditional regime based on land filling was much more the same in all these countries a few decades ago? We sum up these questions by attempting to demonstrate both similitude and differences between national developments and by questioning the very existence of institutional trajectories of national regimes of waste management: is the inter-temporal coherence in the transformation of waste management regimes such that we can really speak of trajectories?

Proceeding to explore these questions is certainly a stimulating exercise in itself. It is also critical in the context of the European single market and for any thinking concerning the development of European policies in the field of waste management. This field is, indeed, considered to be one of the main areas of interaction between environmental and free-trade issues. The question as to whether waste should or should not be submitted to the same rules of movement as any other commodity within the single European market is an important one and not a new one. The introduction of measures favouring the introduction of re-use in packaging, mainly through deposit–refund schemes, is also the source of a free-trade controversy between supporters of preventive measures and the commodities industry. These and other elements have given birth to a consideration of the need for a European waste management policy, with the consequent adoption of directives as early as the seventies.

Most recent attempts at the strong and deliberate harmonization of national regimes in this field (the packaging directive) have encountered difficulties at the negotiation stage. The result obtained has generated a feeling of deception amongst environmentalists and in pro-European circles. Does what we demonstrate in the recent developments in five European countries really justify this deception? It would seem not. Our comparative study may support another view. Through various adjustments achieved in response to difficulties met at the implementation stage by recent reforms, the main differences seem to fade away, unexpectedly bringing regimes closer together than had initially seemed possible. Does it mean that, although an institutional convergence is not directly enforced by binding European rules, some hidden forces impel regimes towards such a convergence? Convergence cannot be finally excluded, but there exist theoretical reasons for not viewing it as an inescapable outcome. Insofar as developments are driven, at least partly, by visions of the future and, specifically, visions of sustainability, the plurality of myths describing the nature of sustainable waste management maintains the possibility that a significant distance will exist between national trajectories in the future.

Concerning the comparative approach chosen, doubt might be expressed concerning the choice of countries and the absence of important ones such as the UK. There is no doubt the choice of five countries out of 15 gives an incomplete picture. However, the five countries chosen have the advantage of representing, proportionally, the diversity of member states. Germany and the Netherlands typify the category of northern, allegedly more environmentally aware, countries. Italy and Greece are opposites; southern countries that are often said to confer a lesser importance on environmental topics than the European average. Finally, France is half way between these two poles, belonging to the former category in some areas and to the latter in others. This, at least, is the common impression these countries give at first sight, but our study shows that this is not always so simple.

The book contains six chapters: five chapters each describing a national regime and a final comparative chapter. Although national chapters are independent and may be read separately, the order in which they appear is not random. The chapter on Germany is so placed in order to introduce, in addition to the particularities of the country, some general characteristics observable in other countries. These tendencies have not appeared in the same way everywhere, nor with the same intensity. Nevertheless in Germany, as in less advanced countries from the standpoint of waste management, such as Greece, they are at the origin of the movement favouring a redefinition of the entire field. The Netherlands, close to the Germans in terms of a preoccupation with the environment, also display the characteristics of a national regime that began to develop 20 years ago. France (Chapter 3) and Italy (Chapter 4) contain regimes which evoke comparatively new experiences, in the sense that institutional mechanisms have been introduced fairly recently, while Greece (Chapter 5) is still in the phase of reflection as to which regulatory mechanism to introduce.

As has already been alluded to, there is no evidence that each country evolves in the same way, in spite of a spontaneous process of erosion of the most extreme differences. Yet, the reading of the five chapters may reveal a panorama of national regimes contemplating different development options within their own institutional trajectories. The synthesizing chapter develops this point by recapitulating in a systematic manner differences in the organization of municipal waste management and reducing these differences to the interplay of a few variables.

Descriptions and analyses presented in this book have been obtained as the first part of a three year European research project on municipal waste management, called SUSTAIN WASTE, that started in 1996 with the support of the European Commission.[3]

Notes

1. Traditionally, the other is water management, whilst a new one is emerging: energy choices as environmental issues, mainly through the case of global warming.
2. This is not to deny any significant differences amongst these European countries, but these differences are far less important than those we observe with Eastern Europe, Africa or South-East Asia, to cite but a few examples
3. European Commission DG XII, Environment and Climate Research Programme (1994–1998): Research Area 4: Human Dimensions of Environmental Change

CHAPTER 1

Municipal waste management in Germany

Lilo Fischer and Ulrich Petschow[1]

Introduction

The discussion in Germany about waste management started in the 1960s. Traditional waste management, however, was basically rooted in overcoming hygiene problems. The Federal Epidemic Act (1900) required public provision to be made for basic requirements (*Daseinsvorsorge*) under the responsibility of local governments. The implementation of public health objectives should be supported by compulsory use of public waste management services free of charge. After World War II, the local self-government guarantee was institutionalized in the Constitutional Law to implement the right of citizens to self-determination.

In Germany there has always been the ability to exclude industrial waste from public waste disposal due to composition or amount, thus leaving this responsibility to the private sector. Although wastes are classified into three categories according to their origins (domestic, commercial and industrial waste) there is a lack of clear cut borders between those types. Despite the evidence that there are dependencies between the developments in the different sectors, the focus is on domestic waste – only referring to the other origins in case of mutual influence. Waste from human settlements is delivered to public waste disposal authorities and includes municipal waste (about 40 million tonnes), industrial waste similar to domestic waste (about 7 million tonnes) and building rubble. In the 1990s, municipal waste includes domestic waste (about 30 million tonnes including bulky refuse) as well as commercial waste similar to domestic waste, city cleaning and markets (about 10 million tonnes).

Driving forces behind the development of a tangible waste disposal regime were factors of dissatisfaction which primarily resulted from the effects of the growing 'throw-away' society. These led to a change in the composition of waste on the one hand and, consequently, to new problems which could not be dealt with under the existing distribution of institutions responsible for the disposal of waste. The local authorities were responsible for waste manage-

N. Buclet and O. Godard (eds.), Municipal Waste Management in Europe, 5–62.

ment, but were, in many ways, not technically or administratively capable of dealing with these problems. The diverse problems of co-operation were especially difficult to overcome in this context. This resulted in many uncontrolled and unsafe landfill sites which were ultimately the main cause of dissatisfaction. Finally, it was through the (*Hochzonung* and) development of new institutions, i.e. through the distribution of tasks at a centralized state level and/or *Länder* level, it has been possible to overcome these shortcomings.

In summary the waste disposal regime was based on the public provision for basic requirements, the obligatory provision of services (*Andienungspflicht*) etc., which ultimately led to the local authorities having a far-reaching responsibility. This was later criticized, however, because the production sector was in the end not responsible for accumulating waste. Production, transport/shipments of waste, and disposal were the responsibility of different types of systems (in this context, the privatization of profits and the externalization of social costs/burdens were a central issue for the environmental movement).

These diverging systems were problems which were initially ignored. This was primarily because as time went on, the waste management represented a moving target. Estimates of waste composition from the beginning of the 1970s are very different from those of the 1990s. Recycling, for example, and/or the sorting of waste streams became normal practice over a long period of time and this was organized by the private sector where glass and paper, for instance, were collected separately by private firms. The change in the composition of waste resulted mainly from the changes in relative costs/prices and different consumer habits. In the end, many substances which had been classified as valuable were now devalued and regarded as waste. Thus, the whole situation of the waste industry was fundamentally changed as increasingly more substances became waste; a direct consequence of the 'throw-away' society. In this respect, the waste industry increasingly became an outlet for producers who had to deal with the environmental impact (primarily waste) of production to a limited extent or not at all. In the long and short term, this has consequences on the waste disposal regime, since under the conditions of the public provision for basic requirements, the state was ultimately responsible for the waste streams without being able to intervene in the causes of waste. The established tendencies for the privatization of waste disposal management today are, at least in principle, aimed at transferring responsibility to the producers of the waste streams. This altered perspective inevitably results in necessary adaptability to the institutional structure of the waste disposal regime which is at present, however, only partially recognizable.

This analysis of the development of the municipal solid waste management in Germany is divided into six sections. Section one starts with the state of municipal solid waste management in the 1960s and the need for federal action. In 1971, the Environmental Programme emphasized the need to shift tasks and responsibilities to a higher level. The additional financial, organizational and technical resources required to establish controlled landfilling, could only be afforded by co-operation and the realization of economies of scale in waste management.

Controlled landfilling became one of the main objectives of the Waste Disposal Act in 1972. The Act already incorporated most of the recommendations made in the Environmental Programme, thus forcing waste disposal authorities to co-operate at the more centralized level of districts. Administrative instructions supported the implementation of uniform standards for landfilling.

Under the influence of the oil crisis and the publication of the Club of Rome report, the Waste Management Programme in 1975 formulated a new imperative: waste avoidance should have precedence over recovery, recovery should have precedence over landfilling. The programme reflected the main issues of discussion until the end of the 1980s (section two). With the Act on Avoidance and Disposal of Waste in 1986, priorities were moved towards environmentally motivated prevention and legally defined recovery and the enlargement of the scope of the waste regime to include wastes for recovery. In practice, however, the elimination of waste remained the dominating option in household waste treatment chosen by Waste Disposal Authorities (WDA), documented and criticized in the report on waste management by the Expertś Committee on Environmental Questions in 1990. Delayed investments and the avoidance of conflicts at a local level caused by the NIMBY syndrome led to a decrease in future capacity. The 'disposal emergency' at the end of the 1980s was mainly caused by the awareness of insufficient capacities for the amounts of waste produced.

The demands on the political agenda to promote and encourage prevention and recycling effectively coincided with a movement in policy style from co-operation towards intervention in the *Töpfer-era.* Discussions focused on packaging waste as a phenomenon of the throw-away society and as a major contributor to the disposal emergency, and on the noxiousness of co-disposal and the long-term safety of landfill sites. Moreover, resistance to the construction of incinerators put pressure on WDAs as well as on industry.

The Packaging Ordinance representing a new approach in municipal solid waste management, developed a new (private) regime for the management of packaging waste and enforced the concept of closed substance cycles. The Closed Substance Cycle and Waste Management Act of 1996 supports both of the tendencies by the reflected intention to privatize commercial waste management and to extend producerś responsibility to other products. The Technical Instruction on Waste from Human Settlements determines costly standards for landfill sites as well as for landfills by the year 2005; moreover, organic waste has to be separated at source. The centrally claimed standards on waste going to landfill sites are, however, very controversial because there is no technical choice: the only state-of-the-art technology to meet the standards on waste is incineration.

The following chapters focus on packaging waste management as an illustrative example for climbing up the prevention ladder. The concept of extended producerś responsibility is thought to incite source reduction as well as efforts towards a higher level of recycling. The NIMBY-syndrome strengthened the position of recycling to a large extent. The strict implementation of closed substance cycles by institutionalizing extended product stewardship will bring about a derogation of the classical principles in waste management such as the proximity and self-sufficiency principle.

The first phase – the need for federal legislation: centralization and standards to overcome environmental and hygiene problems in the 1970s

At the beginning of the 20th century, a Public Health Regime emerged in answer to severe problems caused by private waste management responsibility and the effects on water quality. The enacting of the Federal Epidemic Act in the year 1900 was due to the fact that Prussia, the most important *Land* sited downstream on one of the most important rivers, was affected most by private waste disposal activities. The first chapter describes the further development of the Public Health Regime and the responsibility of municipalities in the 1970s.

The institutional setting in the 1960s

Responsible authorities

Traditionally, local authorities have been held responsible for waste disposal as a task of public provision for basic requirements (*Daseinsvorsorge*) due to the fact that the local environment dictates the degree of danger caused by pollution. After World War II, article 28 II of the Constitutional Law provided a *guarantee of local self-government* to implement the right of citizens to self-determination. Municipalities, however, had to obey the minimum requirements of various Acts[2] focusing on sanitary aspects in order to protect human health.

Instruments

The institutional setting of waste treatment was initially determined mainly by *local ordinances.* Whereas for households, the most widely applied instrument was compulsory usage of waste bins (*Anschluß- und Benutzungszwang*),[3] specific types of waste such as industrial waste could be excluded from public waste treatment. On the basis of the Local Taxation Act, the municipalities were allowed to finance their activities by fees equivalent to the service offered (equivalence principle and cost coverage principle). In 1971, 75% of the population[4] was covered by the system of compulsory usage. Collection was sometimes entrusted to private firms.

Technical choice

Uncontrolled dumping just outside the city limits was common at that time (78% of collected waste).

Factors of dissatisfaction emphasize the need for an institutional change

The recognition of environmental problems

The Interparliamentary Task Force for environmentally sound economic activities[5] and the demands of the Green Charter of Mainau in 1961 showed an increased ability to recognize environmental issues. The early ideas and challenges, such as the protection of the landscape, threatened by numerous landfills, were put on the political agenda during the election campaign of the SPD (Social Democrats) in 1961, and even had a significant impact on environmental discussions during the 1970s. The growing importance of such issues could also be seen in the increasing number of questions in the Federal Parliament and the initiation of legislation concerning environmental problems in several *Länder*.

The recognition of waste disposal problems and creation of central institutions

Due to the unexpected increase in the amounts and composition of wastes (from 1950 to 1961 municipal waste had doubled by weight) and the continuing lack of suitable sites for waste disposal, municipalities as well as industry faced severe difficulties in waste disposal. Moreover, hazards and nuisances caused by 50 000 unauthorized dumps were becoming evident, such as contamination of bodies of water and groundwater, emission of dust and odours, release of toxic gases, fires that were difficult to bring under control, and insect infestations.

The first federal report on the problem of waste disposal in 1963 documented the extent of problems to be solved and made the public aware of them. The parties involved announced their general willingness to co-operate in establishing a body to work out general principles and directives. When, in 1965, the Central Office for Waste Disposal, under the direction of the Federal Health Agency,[6] was established to consult municipalities regarding scientific or organizational questions, the costs of implementation had to be shared equally by the Federal and *Länder* Governments. The resistance of industry and municipalities in contributing to this institution was documented in the second federal report on the problem of waste disposal in 1966. The *Länder* formed a special co-*Länder* working group on waste management (*Länderarbeitsgemeinschaft Abfall* LAGA) in order to confer with each other (and the Federal Government) and to coordinate tasks.

The emphasized need for an institutional change

The technical and organizational measures undertaken were not adequate to achieve regular (and harmless) disposal. The unsatisfactory state of landfilling was mainly caused by a lack of know-how concerning planning, equipment, maintenance and operation of landfill sites.[7] Moreover, traditional solutions at the local level (small areas) as well as national legislation which was oriented mainly towards problems of public hygiene, proved to be insufficient.

The new government coalition of Social Democrats (SPD) and Liberals

(FDP) in 1969 signalled an environmental turning-point with the publication of an action programme to protect nature, including the first proposal for a Federal Waste Act. After responsibility for the environment was shifted from the Ministry of Health to the Ministry of the Interior, the new and enterprising ministerial staff began to work out a far-sighted, and comprehensive plan, reflecting the growing public concern (Table 1).

Table 1. The main ideas of the 1971 Environmental Programme concerning waste

- Re-introduction of waste into the economic cycle;
- Shifting waste disposal authority to a higher level of administration to facilitate efficient and harmless landfilling (ecology through economy)
- To develop waste disposal plans in accordance with regional planning;
- Application of the polluter-pays principle by means of an appropriate fee structure;
- Creation of adequate capacities for disposing of hazardous wastes;
- Financing research and development of alternative waste treatment facilities;
- Directives focusing on the agricultural recovery of sewage sludge and on throw-away packaging.

The creative and forward-looking planning of the ministerial staff was unquestionably the main reason why this programme already included many of the elements of the following legislation.

Spill-overs caused by waste disposal legitimates federal jurisdiction

When a draft law was already in its legislative procedure,[8] the *Länder* tried to impede it by questioning the federal legislative authority for waste disposal. Although they intended to maintain their overall authority, they agreed upon the need for a federal framework supporting minimum requirements to avoid spill-overs between municipalities. Neither the Federal Council nor the newly formed LAGA, both representatives of the *Länder's* interests, could assert their interest in maintaining their authority successfully. The Minister of the Interior insisted successfully on uniform federal regulation when the Mediation Committee[9] was assigned the task of finding a compromise between the conflicting interests concerning the allocation of authority. The constitution was amended to place waste disposal under federal jurisdiction.[10]

The Waste Disposal Act of 1972

The Act reflects the dominant aspect of the 1971 Environmental Programme: preventing any danger to the environment and public health from sanitary landfill sites by enforcement of uniform state-of-the-art standards. The institutional shift of responsibility reflects the Environmental Programme in that equipment, vehicles and facilities which have to comply with hygiene and technical requirements can only be economically efficient if large amounts of waste are supplied to waste disposal authorities (WDAs).

The dual concept of waste defines the scope of the waste regime (Article 1 I)
Regular disposal includes collection, transportation, storage, treatment and disposal in landfills of movable materials that the owner wishes to get rid of (subjective concept of waste) or which could threaten public health (objective concept of waste). Concerning domestic waste, the distinction has no significance, although the definition of waste still remains one of the most controversial issues.

Minimum standards to overcome inactivity (Article 2)
The Act obliged the Länder to protect water, soil, plants, nature and the countryside, and to avoid noxious pollution far exceeding existing requirements.[11] Sanitary landfilling should be achieved according to uniform minimum standards, controlled by compulsory usage of facilities. Requirements concerning adequate sites, size and equipment should be regulated by specification sheets agreed upon by the *Länder*. Facilities which did not even comply with existing provisions (such as those of the Water Act) could be closed immediately.

Responsible authorities (Article 3 II)
WDAs were still responsible for domestic waste, however, the traditionally local task of waste management was made a task of central government by the shift of responsibility to a higher level of administration: the districts.

Instruments
The former compulsory usage of waste bins was amended to compulsory usage of facilities (*Anlagenzwang*) in order to control infrastructural development by uniform state-of-the-art standards (Article 4 II). To implement planning according to regional aspects, a permission procedure for waste-treatment facilities (*Planfeststellungsverfahren*) included implicit proof of environmental compatibility, appropriate sites, and the objective need for a new facility (Article 7 I). To facilitate the assessment of necessary future investments, *Länder* were obliged to draw up waste disposal plans which were compulsory for WDAs.

Moreover, the Federal Government is authorized (given the consent of the Federal Council) to issue statutory regulations regarding beverage packaging, in case disproportionately high landfilling costs arise (Art. 14). This already represented a potential power to intervene in the packaging market by applying economic measures. However, the Federal Government suspended the idea of intervention when respective industries came to a voluntary agreement in 1977.

Results obtained – the implementation of the Act

Delays due to conflicts while shifting waste disposal authority
Political disagreements, difficulties and conflicts during the 1970s caused delays of up to 10 years in achieving the intended results.[12]

The local self-government guarantee creates uncertainty about the legitimacy of the shift. The flood of legal suits by several municipalities demonstrated their desire to hold on to their responsibilities, thus creating uncertainty in the early 1970s as to who should decide on the necessary investments for sanitary landfilling. Their arguments were based on the constitutional principle of subsidiarity with respect to local affairs. The conflict was settled by the ruling known as the *Rastede decision* in 1979. The ruling of the Federal Administrative Court unequivocally shifted the responsibility to the districts. On the one hand, the ruling was based on the objective of the Act – harmless waste disposal could only be achieved by enlarging the areas of organizational responsibility – and on the other hand, the Court stressed the need for adapting public administration to economic development, to make it more efficient.

The difficult implementation of waste disposal plans. Until 1979 the situation was quite unclear and supervising agencies responsible for the implementation and conformance to waste disposal plans had to bargain with local authorities. The requirement to plan according to regional concerns (interpreted as the first derogation of the proximity principle) led to an institutionalized conflict between the regional and the local level. They co-operated by finding compromises with respect to closures and the quality of a revised waste management which, in general, led to incremental improvements of the waste treatment infrastructure. The most important qualitative improvements took place due to pressure to settle the conflicts arising between WDAs and the locally based citizens' action groups campaigning against dust and odour emissions from existing facilities.

At this point, the role of citizens' action groups who tried to prevent the establishment of waste disposal facilities in their neighbourhood should be mentioned. It was the local politicians who had to bear the costs (political) of the centrally defined objectives because the feed-back and communication system between the executing level and level of political programming was disrupted, due to the decrease in local rights to participate in solving waste management problems.

Organizational change
Finally, WDA could be re-transferred to municipalities if desired. Due to the local self-government guarantee, there was no alternative than to offer at least this possibility of re-transferring responsibility for collection and transportation. The *Länder*, having jurisdiction at least over the possible administration of responsibilities at different stages of waste disposal, made use of this possible

re-transfer and authorized districts to decide about it. The districts were made responsible at least for landfilling and other disposal treatments. However, in most cases, they also organised themselves concerning collection and transportation services. The possibility of WDAs entrusting third parties (such as private companies) was mainly used by the districts, which were confronted with these new tasks by the shift of activities to this higher level of administration. In contrast, in the cities, the well established municipal administration continued to carry out waste management as a direct local government activity (*Regiebetriebe*).

Techniques of treatment

Landfilling. The main success of the Act can be seen in the overall reduction of the 50 000 uncontrolled dumps (including only 130 sanitary landfill sites) due to a rigorous regrouping of neighbouring activities by a *local government reorganization* initiated in 1967 in order to improve the efficiency of the administrative structures. The formation of larger units of local administration resulted in reducing the number of municipalities from 24 300 in 1965 to 8400 in 1982. In 1982, all households were involved in the system of collection and the 50 000 dumps had been reduced to 439 sanitary landfill sites. Fifty-two transfer stations were established in 1975 to realise economies of scale by gathering waste for further transportation. Moreover, transport vehicles and waste bins grew bigger, making collection more efficient. Table 2 indicates the outstanding difference between areas of waste density in cities and other areas due to realized economies of scale.

Table 2. The structure of costs in 1977 (ECU/tonne)

	Cities	Other areas
Collection costs	19.12	38.90
Transportation costs	(10 km)	(15 km)
	7.73	11.25
Landfilling costs etc.	10.00	10.00
Total costs of disposal	36.85	60.15

Source: Doedens (1982), p. 26.

The municipalities' superficial search for alternatives resulted in traditional landfilling. In general, the respective building committees were not interested in environmental concerns, and the local parties responsible were not able to decide or solve either technical or environmental problems.

The practice of incineration is marked by variability. The number of incinerators increased between 1965 and 1977 when 36 facilities were built. Since 1977 only nine additional incinerators have been established. The percentage of inhabitants served by incinerators increased from 4.2% in 1965 to 14.4% in 1970 and 22% in 1975.

One of the most important instruments for controlling infrastructural development were the *golden restraints*: golden because *Länder* contribute to local investments, restraints because they demand conditions in return for their financial contributions. Consequently, they provide incentives for the municipalities to act according to wider preferences. This constrained financing partly explains why there were such different paths taken within infrastructural development according to regionally preferred technical options.

Bavaria especially preferred incineration as a technological option: 20 incinerators are situated there today[13] and the percentage of inhabitants served by incinerators is far above the average in Germany as a whole. As in most *Länder*, Bavaria made districts responsible, but gave them the option to re-transfer the shifted waste disposal authority. In 1973 the districts were given 5 years to take over traditionally local responsibilities from the 8000 municipalities. In 1978 only 96 responsible units were left in Bavaria, not least due to the local government reorganization which was initiated in 1967. The political desire to invest in incinerators was reinforced by the incentive of the golden restraints. Another reason for increased incineration is a lack of space in Nordrhein-Westfalen for example. In the other *Länder*, incineration remained a marginal option.

Composting as a neglected option. Whereas 44 composting facilities had been planned in 1975, composting remained a neglected option due to costs[14] and the difficulties in establishing markets for the compost. A considerable increase was observed in the late 1980s.

Conclusion: legitimate shift creates path dependency

Municipalities demonstrated finally the need to create larger units, not only due to the organizational, technical, and financial resources required, but also because of their inability to work in an economically efficient and/or environmentally sound manner. The intention of handing over responsibility to districts was to improve the cost-effective operation of facilities as well as to use the shift to leave old methods.

Closures of landfill sites occurred due to the rigorous regrouping of neighbouring activities with the local governmental reorganization in the 1970s. The golden restraints were applied to encourage *Länders'* preferences in waste treatment. Decisions at regional levels for or against incineration determined future interests in the making of regulations at the central legislative level in favour of or against this technical choice: the *Länder* which decided to build incineration units early on are applying the pretreatment today, whereas the others do not.

The second phase – factors of dissatisfaction favouring a radical change: discussions to overcome the disposal emergency in the 1980s

Whereas the first phase had focused on the removal of the annoying waste created by a consumer society, in the second phase the dangerous potential of these residues, especially for future generations, was entering the public consciousness. The main issues were the problem of concentrated pollution hotspots (*Altlasten*) and waste tourism (shipping waste across long distances and even exporting it, to evade stricter regulations).

Recognition of and reaction to new problems during the 1970s

Oil crisis and limits to growth

Growing awareness of the dissipation of natural resources and the decreasing ability of the environment to absorb toxins began with the famous report of the Club of Rome *Limits to Growth* published in 1972.

The oil crisis in 1973 coincided with this evolving awareness, and the idea of recovery has since gained increasing significance in Germany. Economic reasons, such as the use of waste as secondary material in order to diminish dependence on imported resources, took on greater importance as a result of this crisis.

First legislative reaction: the order to recycle addresses industrial facilities

The *Federal Pollution Control Act* of 1974 might be characterized as the first reaction to the oil crisis with respect to regulations concerning waste. The motivation for an institutionalized order to recycle (Article 5 I Nr. 3) not only consisted of the intention to reuse waste as a means of resource conservation, but also to relieve landfill sites and decrease dependence on imported resources for economic activities. The Act only affects decisions in the production sector, as the order to recycle refers only to facilities licensed in accordance with the requirements of the Federal Pollution Control Act. However, the idea gained importance for the development of the waste management regime on the whole.

Basic ideas for a new approach in waste management: the Waste Management Programme in 1975

In 1975, the Waste Management Programme (WMP 75), as the first detailed and systematic analysis of the whole waste situation, was developed by the Federal Government in co-operation with science and industry (Table 3). Still under the influence of the oil crisis and the *Limits to Growth* report, a new imperative, comparable to *Landsink*'s idea of a prevention ladder, was formulated: Waste avoidance has precedence over waste recovery, waste recovery has precedence over landfilling.

The WMP 75 indeed showed ways of increasing sustainability, but they were still not mandatory. As a direct consequence, the *Environmental Statistics Act* of 1975 supplied the basis for further inquiries. However, it took a decade of

Table 3. Objectives of the Waste Management Programme

Reduction of waste in production and consumption by:

- Reduction of industrial waste;
- Introduction of environmentally sound production technology;
- Examination of material assignment with regard to its adequacy;
- An increase in the durability and repeated use of products.

Increasing the use of waste by

- Waste recycling;
- Profiting from its energy content.

Harmless disposal of waste.

Attribution of costs according to the polluter-pays principle.

According to the WMP 75, the improvement of the following aspects was to be promoted:

- Statistics as a basis for planning;
- R & D with respect to recycling and treatment procedures;
- Development of markets for recycled products;
- Alteration of products (prevention principle instead of end-of-pipe technologies);
- Organization of waste management;
- Waste legislation.

discussions to move priorities from elimination towards recovery. The waste system was extended to include the encouragement of avoidance and recovery through the 1986 *Waste Act on Avoidance and Disposal.* However, only the disposal emergency at the end of the 1980s reinforced priorities to be switched effectively and put into practice.

Amendments of the 1972 Act responding to several abuses

Several abuses, usually exposed by the ecological movement and documented by the mass media, forced those in power to react by amending the 1972 Act. The problems arose due to a loophole in the law concerning public responsibility for industrial waste treatment: waste which could not be treated together with domestic waste, due to its composition or amount, could be excluded by ordinances of WDAs. The re-transfer of responsibility led to disastrous future problems such as pollution concentrations, and exports of toxic wastes because the market was not able to supply the necessary infrastructure to treat the wastes. Therefore the first amendment of the Waste Act in 1976 focused on improvements of control. The second amendment of the Waste Act in 1982 was intended to bring about harmless recovery of sewage sludge, and improvements in the administrative law. A third amendment in 1985, with the objective to improve the control of cross-border waste transportation, was made to emphasize the willingness to implement the normative proximity principle of mutual respect as a result of accumulating scandals and waste tourism.

The disposal emergency

The disposal emergency represented a situation where the landfilling capacities declined in most regions and the planning of new facilities was avoided or delayed due to conflicts. Consequently, the 1980s were overshadowed by the disposal emergency when facilities could not be built at a rate to match the growing volume of waste. Awareness of the disposal emergency was accentuated by international discussions and global conventions which impeded disposal at sea and emphasized the problems and limits of landfilling activities, especially in Germany.

Waste avalanche

There was an increase in the amount of municipal solid waste by weight between 1960 and 1970 and a considerable increase in commercial waste (Figure 1).[15] The waste avalanche in the 1980s, however, mainly occurred due to the increased *volume of waste* (in contrast to weight).[16] A significant change in composition of municipal waste occured when inert waste such as ashes was increasingly replaced by paper starting at the beginning of 1970s. With this changing composition, there was not only the need to revise landfilling standards but also the possibility of reduction at source.

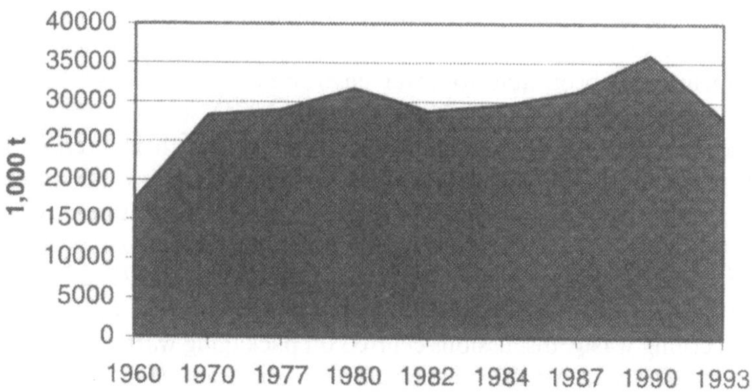

Figure 1. Municipal solid waste arising in Germany from 1960 until 1993.
Source: Statistisches Bundesamt, Fachserie Umweltschutz, Reihe 1.1., Serie 19, 1998

In particular, the sudden increase in packaging materials was seen as one of the main contributors to the increased scarcity of landfill sites in the 1980s. The highest growth rates occurred in the plastic industry. Paper had to suffer most from this increase and the proportions of other materials remained constant. The packaging industry was used to growth rates far above the average growth of the gross national product: packaging production increased from 3.8 million tonnes in 1960 to 9.3 million tonnes in 1974.[17] The break-down in 1975 and again in the time between 1980 and 1983 was due to the oil crisis with increased prices on virgin materials.

The 'not in my backyard syndrome'

The increasing reluctance to accept any kind of waste disposal facility within the population (known as the not in my backyard syndrome) led to the problem that planning could not be enforced. The two main reasons for this were first the diminishing social acceptability to bear the environmental costs caused by the misuse of landfill sites and incinerators as further centralization took place and second, the reluctance of municipal decision-makers to tackle things in order to avoid conflicts with voters at a local level. Local resistance met with the restricted possibilities to act at non-central levels: municipalities, districts and *Länder* had neither legal nor actual competence to influence waste amounts effectively; these corporations view the discrepancies as political costs.[18]

Focus of discussions in the 1980s to overcome the disposal emergency

As a consequence of the centralization and poor implementation in the first years of the Waste Disposal Act, citizens' initiatives (NGOs) concentrated more and more on the protection of their interests at the local level. Since jurisdiction had not been able to cover environmental concerns, the public consent seemed to be endangered. The increase in NGOs as a starting point of the ecology movement in Germany reflected the fact that their interests were hardly represented politically. A discussion began on how the interests concerning public environmental commodities might be institutionalized. The Green Party evolved from these citizens' initiatives (Federal Association of Citizens' Initiatives for Environmental Concerns founded in 1972) and entered traditional politics when they were voted into parliament as a fourth party in 1983. Concerning waste, discussions centred on packaging waste, safe landfills and clean incineration. All these aspects led to important changes within the waste management regime and can be interpreted as an outcome of the strong pressure caused by environmental movement in Germany.

Growing awareness about problems bound to incineration and codisposal

German opposition to incineration. In the 1980s, the catastrophe of Seveso (and especially the uncertainty of the whereabouts of the dioxin waste from Seveso) caused the German population to be very sensitive toward the issue of waste disposal. Waste incineration as a possible solution to the disposal emergency brought about even more local resistance since the population was aware of the risks involved. When local conflicts and high investment costs had hindered the provision of necessary facilities to dispose of the waste, the German waste crisis reached its peak in 1990. The environmental movement denounced thermal treatment (interpreted as an option to eliminate only harmful substances) because of toxic emissions, and especially because there is no chance of recovering all the energy needed for the production of the product. They argued that incineration is a waste of energy and scarce resources, and that a move toward waste avoidance could be only reached by the consequent closure of waste treatment facilities.

The leachate problem of landfill sites. A landfill site which had been closed in 1979 (Hamburg, Georgswerder) was found to be leaking in 1983 because toxic waste had been embedded within municipal waste. Since many harmful substances, especially dioxins had contaminated the soil, it had to be cleaned up at a cost of several hundred millions of ECU. This event emphasized the need to find alternatives to landfilling which cause unpredictable harm and therefore higher costs in the long-term. The experience with leachate brought about a rather broad discussion over the safety standards of landfill sites in Germany. Although some technical progress could be made, i.e. by sealing off landfill sites, collecting and purifying seepage water, as well as recovering and using landfill gases, experts complained that there was still no long-term security of landfill sites. In 1990 the Council of Environmental Experts published a special report on waste management in which it was emphasized that residual waste should only be dumped when its consistency was similar to that of the earth's crust. The Technical Instruction on Waste from Human Settlements (TASi) resulted mainly from the discussion about the waste avalanche in the eighties and the growing awareness of waste disposal emanating from serious leachate problems.

Switching priorities in waste management towards recovery

The conflict between the ecological need to restrict landfilling to a few adequate sites on the one hand, and the proximity principle and the self-sufficiency principle aiming at the congruity of cost-bearers and beneficiaries (requiring a decentral structure of waste disposal) on the other brought about changed priorities in waste treatment.

Switching priorities on paper: the 1986 Act on Avoidance and Disposal of Waste. Due to public and political pressure, the only alternative was to establish a new hierarchy of objectives as described in the WMP 75. The emphasis of the Act on Waste Avoidance and Disposal reiterated those early ideas to prevent any accumulation of pollutants, and therefore of further environmental damage, and to adopt a new hierarchy to promote recovery as opposed to disposal in view of growing awareness that landfill sites threaten future generations to an unforeseeable extent (interpreted as an intergenerational proximity with respect to external effects caused by waste generation).

The basic changes of the 1986 Act were:

(1) The legal definition was no longer appropriate (Article 1 I). The definition of waste, determining the scope of the existing waste management regime, was enlarged to clarify uncertainty about what is meant by waste or a commodity, depending on the owner's reason for getting rid of it. The ambiguity of the definition 'waste' was especially a problem when public health concerns were affected (objective waste). As a result, movable assets, even in the case of intended recovery, had to be considered as waste until reintroduced into the economic cycle.

(2) The new hierarchy in waste management (Article 1 a). The phrasing chosen, stating the new hierarchy of objectives, did not possess a legal character.[19] Whereas the priority of waste avoidance (quantitative and qualitative aspects) is only based on an environmental, not legal, priority, recovery serves as a second-best solution in waste management policy. According to Article 1 II, recovery includes the use of the energy content of waste, reduction of weight and volume, and the reduction of noxious substances to reduce the costs for the final step of landfilling. Due to its greater compatibility with environmental concerns, experts, the opposition in Parliament, and the public demanded that recycling took priority over energy recovery. However, it was not specified in the Act.

(3) The central instrument: Federal legislator is authorized to issue ordinances with respect to priority waste. With the amendment of article 14 of the 1972 Act (referring only to beverage packaging), the conditions to intervene were facilitated and enlarged to include products in general. An extended power to intervene in markets was now endorsed:

- Article 14 I aimed at an improved management of hazardous substances within municipal waste streams. The instruments provided included obligatory labelling, or the separation of these materials, obligations to return or charge deposits on items, and bans on specified materials.
- Article 14 II institutionalized the hierarchy of negotiating and intervention powers. Before intervention takes place, the Federal Government has to specify source reduction or recycling objectives. Only if these goals are not achieved may the Government, especially in the case of packaging, order even stricter requirements than are described in Article 14 I of the Act.

The priority of recovery in practice. The main problem in enforcing the priority of recovery effectively lies in the unclear legal expressions used describing prior conditions.[20] Therefore, the federal Act tended to set a theoretical standard rather than practical effects. Thus, elimination was still the dominant option in waste management in practice.

Problem-solving capacity is inconsistent with institutions affected by the problems. Due to the principle that federal jurisdiction overrides *Land* jurisdiction, *Länder* could neither specify further regulations, nor require any other hierarchy of objectives for waste treatment. Although communities, districts and *Länder* as the corporate bodies with planning and executive powers were responsible for solving the conflicts, they had neither legal nor real power to influence waste amounts.[21] The local conflicts about new facilities which non-central political institutions had to face were perceived in fact as political costs. The growing awareness of scarce future capacities in German municipalities pushed forward the reform of federal waste management policy: together, they started to make demands on the Federal Government to switch priorities in waste management by enacting ordinances according to article 14, and complained that regulations were only planned and never implemented.[22]

Efforts by WDAs. The priority of recovery also applies to WDAs (Article 3 II, clause 3). A new trend starting in the mid-1980s was to separate waste at source in order to improve the possibility of recycling, since the only way to overcome the conflicts described was to avoid and to recycle waste in order to gain acceptance from the local population. The WDAs were the main driving force in establishing a system to collect separate materials on a voluntary basis such as paper, glass and organic waste, as well as other priority waste, e.g. batteries, supported by local campaigns to improve public acceptance.

Focus: beverage packaging and adequacy of voluntary agreements

The discussions arising from the disposal emergency focused on packaging waste since it took up a significant proportion of the total municipal waste (50% of the total volume and one-third of the total weight) and due to its symbolic character of the consumer society.[23]

Short history of voluntary agreements. After assuming that co-operation with industry would succeed, the Federal Government made no use of the authorization to issue statutory regulations on packaging under Article 14. Facing decreasing future capacities and disproportionately high costs caused by beverage packaging, the Government announced interventions in 1976 in case of further increases. Under this new pressure, in 1977, the first mutual agreement between the respective Ministry of Interior and the head organizations of the beverage, packaging, glass, and steel industries as well as the retail sector, scrap metal industry and individual big enterprises was based on the bargaining institutionalized by article 14. The Federal Government promised to suspend the idea of an ordinance, industry agreed to stabilize the amount of refillables, to not introduce large plastic bottles and to increase recycling of glass and tin plate. The 'voluntary' agreement only succeeded where economic interests were involved i.e. recycling of glass and tin plate. The main objective, however, to stabilize the proportion of refillables failed. A second voluntary agreement in 1982, to stabilize the system of refillables, also failed.[25]

The history of voluntary agreements demonstrated the need for further measures to intervene in targeting refillable quotas as a means of source reduction: the trend towards non-returnable packaging was induced by further concentration processes in the beverage industry and retail sector. With the move from a non-central structure of provision with small markets and short transport distances to larger markets, the economic efficiency of refillables decreased. The voluntary agreements could hardly stop this development.

The PET/Deposit Ordinance in 1988. With the enacting of the European Directive on Packaging for Liquid Food (85/339/EEC), the Member States had to take measures aimed at reducing volume and weight whilst complying with the determinations of free trade (Article 30 of the Maastricht Treaty). In order to comply with the Directive and to react to the violation of the agreement in 1977 by Coca-Cola which used large volume plastic bottles, the Federal Government enacted the PET or Deposit Ordinance in 1988: disposable plastic bottles carried a deposit levy of 0.25 ECU and had to be taken back by retailers for reuse or recycling. Since retailers were not willing to establish a second deposit–refund system in addition to the system of refillables which already existed, they played a significant role in the disappearance of the PET and PVC bottles from German markets. Although the ordinance was regarded as a great success within Germany, the European Commission evaluated it as discriminating against foreign suppliers. Furthermore, the Commission rejected the special regulation concerning plastic packaging whereas the European Directive referred to all types of beverage packaging.[26] The crucial point of the Ordinance was to use retailers as a strategic factor to enforce political objectives.

The intervention already signalled the willingness to put an end to voluntary co-operation and played a significant role when other voluntary agreements were drafted dealing with the recycling concepts of fluorescent tubes, refrigerators, batteries, and CFCs.

Conclusion: recommendations of the Expert Committee reflect dissatisfaction

In 1990, the Expert Committee on Environmental Questions (*Sachverständigenrat für Umweltfragen* SRU) published a special study on waste management which severely criticized current waste management policy. According to this study, German waste management was still focused on elimination modes instead of encouraging recovery and avoidance. Amounts of waste as well as their harmfulness have been growing continuously, and last but not least, the problem of waste tourism still needs to be solved. All in all, the SRU criticized the legal regulations as insufficient and too diffuse to be effective.

A crucial point in the discussions of the 1990s has been the hierarchy of waste management aims which are not appropriate to achieve real effects despite the pressure put on the regime by the disposal emergency which has mainly been caused by the NIMBY syndrome and the waste avalanche (accompanied by waste scandals). The regime's development concentrates on severe problems of undercapacities and therefore rising disposal prices, as well as towards evolving efforts of WDAs to recycle some waste in answer to local conflicts. The SRU gave a clear-cut statement in favour of giving priority to recycling.

Although political intentions did not cease to give priority to avoidance of waste, legal embodiment of the concept remains difficult.[27]

The SRU finally suggested measures it considered appropriate for a shift from the elimination-oriented throw-away society towards waste avoidance and recycling. In their opinion, human nature as well as market conditions would both withstand a change towards a more preventive waste management policy. Hence, they recommended use of the power to apply administrative rules, e.g. raising disposal costs, or preventive measures, such as legal regulations for product design and production processes or even the phasing out of certain industrial sectors completely (e.g. chlorine industry).

Some of the suggestions of the SRU were taken up and implemented by the Federal Government and the Federal Ministry of Environment. This resulted in several statutory ordinances on hazardous waste in 1990/91, the Technical Instruction on Waste in 1990 (proof of disposal, definition of disposal paths, more concrete concepts for recycling), the Packaging Ordinance in 1991, the Technical Instruction on Waste from Human Settlements in 1993, and finally, the Closed Substance and Waste Management Act in 1996.

Moreover, German legislation could no longer do without a harmonization with EU legislation. The German waste concept did not comply with the European Directive on Waste (75/442/EEC, amended 91/156/EEC). Within the German waste management regime there was an incentive to declare waste as secondary products in order to avoid the stringent and costly German standards. As a result, the supervision of waste transportation crossing borders failed because secondaries were not classified as waste.

The third phase: the rise of a new approach in the 1990s – focusing on the issue of packaging waste

Regulatory changes

The change in regulatory style from co-operation towards a policy of intervention
In the past the Ministry had pursued a distinctive co-operative policy style between bargaining, the solution of factual problems and overcoming conflicts of distribution. The move towards intervention took place as a reaction to the unfulfilled commitments described.

The institutionalization of a Ministry on Environmental Affairs in 1986. Not least as a consequence of the Chernobyl catastrophe and the federal election at hand, the Ministry of Environment, Nature Conservation and Nuclear Safety was institutionalized in 1986. In 1987, Mr. Töpfer became Minister. Due to his special knowledge and assumed authenticity, environmental and industrial associations began establishing a cross border information network on environmental issues. Negotiations took place in institutionalized public hearings which were used by the Ministry to find compromises between the conflicting parties.

The determination of objectives in 1989/1990. The normative policy of co-operation had already been institutionalized with the Waste Disposal Act 1972. The policy of co-operation was given first priority: with the precondition, before enacting an ordinance, to determine objectives as politically defined targets which have to be implemented voluntarily. As long as industry, however, had no need to fear that a deterrent would be applied when targets are not reached, voluntary collective action cannot be expected.

The success of the PET Ordinance from 1988 and public pressure[28] encouraged the Environmental Minister for further objectives in April 1989:[29]

- the announcement of interventions and agreement on prohibitions,
- standardizing of a system for refillables,
- different, rather high, quotas for refillables,[30]
- an increased recycling of non-returnable beverage packaging; the beverage industry as a whole should work out a recovery scheme for aluminium, composites and plastics within a period of 6 months.

In particular the specific objectives for the plastic industry and the short time span for enforcement demonstrate that the failure of voluntary initiatives had already been taken into account in order to legitimize further interventions.

Economic instruments in German environmental policy

In Germany there has been much discussion about the advantages of economic measures in contrast to command and control. The proposal of the Federal Environmental Agency (*Umweltbundesamt*, UBA) at the beginning of the 1980s to levy a tax on packaging, also favoured by retailers, had been rejected due to the fact that the burden of the tax (and thus effectiveness) could not be controlled by state governance.[31] On the other hand, the uncontrolled passing on of the costs led to the preference of a packaging tax on the part of retailers. A legal duty for retailers to use refillables as opposed to non-returnable packaging (called the *lex aldi*) also failed in the legislative procedure.[32] The application of other economic measures, e.g. a licensed amount of non-returnable packaging, is restricted by German Constitutional Law. Facing the acute disposal emergency, the challenge for the Environmental Ministry reached its peak:[33]

- to relieve the public sector of waste treatment, the WDAs;

- to assign the know-how of producers and waste generators;
- to vote for the application of the polluter pays principle in accordance with equity;
- to institutionalize a short-term solution with immediate effects;
- to shift responsibility to the industry.

The purpose of the handing over of responsibility for the impact of economic activity was the intended internalization of external effects. The German approach to institutionalization of the polluter pays principle was that the waste producers were to bear the responsibility for overcoming waste problems.

Since the German political system is pre-structured by command and control, economic measures have to be incorporated into the given structure incrementally or combined with interventions. The Packaging Ordinance and the dual system (DSD) represent such a combination of command and control and economic incentives: recycling quotas had been fixed authoritatively at a high level and the license fees to finance the recycling should result in waste avoidance. The requirement to separate packaging waste at source opened the possibility to establish a dual system. In contrast to other Member States, the responsibility for packaging waste management was given completely into private hands, institutionalized by a total-cost-system in contrast to the shared responsibility principle which is mostly applied throughout Europe. The total-cost system can be interpreted as the economic mechanism to enforce the concept of an extended producer's responsibility.

Legislation

The Packaging Ordinance

The Packaging Ordinance (1991) is said to be a turning-point in German consensus-oriented corporatism, institutionalizing the German idea of closed substance cycles by a top-down approach in contrast to the system of covenants in the Netherlands, for example.

The decision process. With the *Lambsdorff initiative* (FDP representing industry's interests), discussions on a dual system in MSW management started.[34] The main intention of the first draft for a Packaging Ordinance (*Töpfer-draft* April 1990) was to force the retail sector to take packaging back. It was the intention to use their powerful interim position between production and consumption as a gate-keeper. In the resulting hearings in 1990, industry managed to accomplish exceptions allowing them to avoid more stringent individual obligations. The Task Force on Packaging and Environment (AGVU), industrial associations and waste management branch worked out a concept based on the idea of a dual system: private industry should be responsible for recovery of packaging waste, WDAs should only be responsible for other household wastes. Due to the pressure from industry, the governing coalition passed a very lenient ordinance at the end of 1990. Environmental associations and experts of the Federal Environmental Agency (UBA) criticized the fact that waste avoidance was not taken into account. The *Länder*, however, managed to hinder the *corporatistic* concept: the Federal Council as the second legislative power rejected the ordinance, mainly because of equating energy recovery and material recycling. The necessary approval was given in April 1991 when compulsory material recycling was stipulated and the stabilization of a set proportion of refillables as a condition for the DSD was agreed upon. The result was not only due to the strong interests of the *Länder*, but also due to the political position of the Social Democrats in the Federal Council and the special political importance of Bavaria (CSU).

Thus, the *Länder* became the driving force for the stringent regulations which had been enacted finally. The leeway for their action, however, was only restricted by Art. 100 a IV of the Maastricht Treaty specifying that measures should not serve as obstacles to free trade. Whereas the SPD-governed *Länder* as well as Baden-Württemberg and Bavaria warned against complying too early and preferred a risky strategy with strong regulations and hard bargaining with Member States of the European Union, the Environmental Ministry pursued a strategy of risk avoidance and used the European Commission as a scapegoat to defend even more stringent regulations demanded by *Länder* concerning quotas for refillables.

Objectives: avoidance as a concept at the end-of-pipe. The Packaging Ordinance aims at avoiding packaging and prescribes recycling rates for each material, if not reusable (article 1). Whereas for transport and secondary packaging (articles 4 and 5), distributors are individually responsible for taking back, reusing or recycling, the Ordinance makes a decisive exception to these individual obligations in the case of sales packaging: The suspension clause (article 6 III) provided exoneration from this duty by creating a privately operated recovery scheme covering all households, thus establishing a system that worked alongside the traditional public MSW treatment. The exception, however, must comply with the public system and with the quotas determined: a mandatory deposit will be levied on non-returnables in case of failing to reach the quota for refillables of 72%;[35] if recycling quotas are not met, the deterrent of individual obligations to take packaging waste back with retailers will be applied (Table 4). Thus, there is an inherent aim of the privately operated system to meet the quotas serving as a trigger to enforce deterrents.

The Annex of the Ordinance explicitly defines quotas of collection and sorting referring to weight and sales packaging use as a total; the final recycling quotas are obtained by multiplying the two.

Table 4. Resulting recycling quotas (percentages of used sales packaging)

	1993	1995
Glass	42	72
Paper	26	72
Tin plate	18	72
Aluminium	18	64
Plastics	9	64
Composites	6	64

The organization of the regime – new participants in packaging waste management.

The development of organization. As the retail sector has been used as a strategic factor, it has become the main driving force in the implementation process: to avoid becoming the national 'rubbish bin', the retail sector will function as a gate-keeper due to its powerful interim position. The main objective of the Dual System Germany (Duales System Deutschland GmbH DSD) was therefore to be exonerated from their individual obligation. ALDI plc and Edeka Inc. as important retailers in food supply[36] were among the first promoters to establish the DSD in September 1990. The establishment of the DSD had been supported by the tin plate and aluminium industry, as well as the Federal Association of German Waste Disposal Business (BDE). The 400 enterprises participating in the foundation of the DSD became the Eco-cartel.[37] For waste management, the prospects of developing markets such as new fields of activities (recycling) led to their willingness to co-operate with the DSD. It took over investment risks and thus determined their future interests.

The plastic industry held an outsider position during the process of establishment, unwilling to give any guarantees on the recovery of collected plastic, in contrast to other respective industries that agreed upon the concept of the DSD. The BDI as well as DIHT, as the two most important industrial associations and co-ordinators of bargaining processes, announced the enforcement of the concept against the interests of the plastic industry emphasizing the macroeconomic interest involved in its success.[38]

Three main factors exerted pressure on the plastic industry to adapt to the system:[39]

- individual duties could be restricted to particular materials (§ 6 IV Pack Ord), therefore coalitions without the plastic industry were possible;
- being confronted with the Minister and the powerful industrial associations, the plastic industry saw their public reputation endangered regarding their ability and willingness to behave according to public health concerns;

- in practice, the 'green dot' became an insider–outsider label[40] since retailers agreed upon the green dot as a criterion for the distribution of products (pull-effect).

The working of DSD. The DSD made contracts with a duration of about 10 years with private and public disposal enterprises concerning collection and sorting achievements and prices. Agreements were drawn up with WDAs to integrate the existing (public) systems, e.g. collection systems for glass and paper. The light packaging material (aluminium, tin plate, plastic, and composites) had to be separated at source into the special yellow bin or bag (provided for by the system of compulsory usage for households). Sorters have to meet some technical minimum requirements concerning the input for recycling, which are defined for each material within the contracts between the DSD and the disposal branch. Non-recyclables are retransferred to WDAs for disposal (incineration or landfilling). In return, WDAs benefit from financial flows of the dual system. To ensure that sorted materials will be recycled, global acceptance and recycling guarantees had to be given to the DSD on behalf of different associations.[41] The tasks of guarantors include:

- organization of transport and distribution;
- contracts with recyclers;
- distribution of costs for recycling;
- taking on risks concerning returns from the sale of secondary materials.

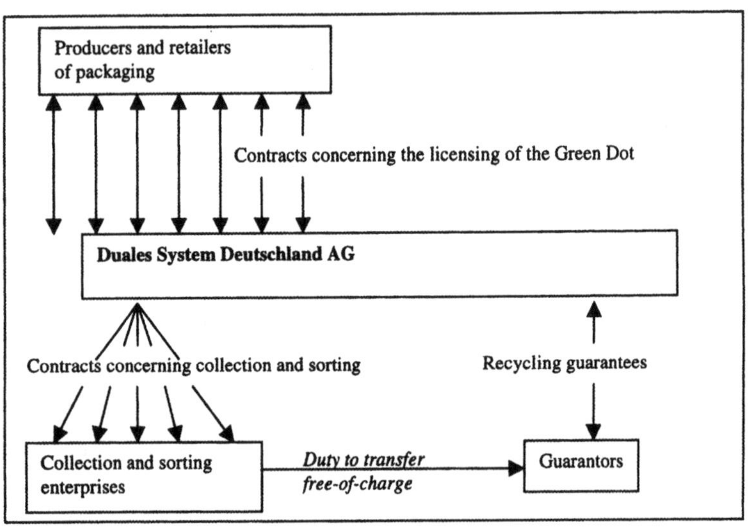

Figure 2. The working of DSD – a broad overview

The whole system of collection and sorting is financed by the levying of license fees known as the green dot. The costs of recycling can be covered by returns of sale of secondary materials or substitution of primaries. In other cases, additional costs have to be shared by the different packaging producers concerned.[42]

Economic interests in favour of high quotas. The disposal branch is one of the few which expanded in the last years at a considerable high growth rate. Those from the disposal branch active in secondary markets benefit from the high and stable packaging waste flows from Germany. They were enabled to strengthen their position in those markets.

Individual industries such as steel production benefit from the high collection quotas because of economic savings by substituting primaries, except for the plastic industry. Since steel mills substitute fuel for mixed plastic packaging, however, the worries about plastic recovery have come to an end. Although the plastic industry faces higher costs, it is better off due to increasing returns: the acceptance of plastic as an alternative to other materials has considerably increased, since the recovery system has been established. Thus, the plastic industry has also benefited from the high quotas. Moreover, there is a great interest in the German solution for plastic recycling (patent of DSD) in Asia.

This illustrates that there are many economic interests supporting the ecological idea of closed substance cycles. The board of directors reflects the interests of the DSD consisting of one-third retailers, one-third disposal branch and one-third packaging industry. They define the conditions of the privately operated system.

The Closed Substance Cycle Waste Management Act of 1996 (KrW-/AbfG)
In comparison to the Act from 1986, the KrW-/AbfG is marked by at least two altered approaches:

- At the legal level: extension of the definition of waste, in conformity with European law designed to overcome the problems of a subjective versus an objective understanding of waste.
- At the executive level: the decision to define a hierarchy of source reduction, recovery and disposal in the context of a comprehensive product stewardship as a step up the prevention ladder prioritizing private waste management in terms of responsibility.

Definition of waste and extension of the concept of waste. The definition of waste has been extended considerably to include all materials and items which are no longer products or whose original purpose no longer exists and cannot be reused in another way. The key question in the 1986 Act of what is waste and what is a product, turns to the key issue in the 1996 Act of what is waste for recovery and what is waste for disposal. The distinction meant that waste for recovery is put under the private regime and waste for disposal is put under the public regime.

The hierarchy of objectives. The avoidance of waste by minimizing its amount and harmfulness now at last took priority over recovery of waste. However, avoidance only takes priority if statutory ordinances have been issued, now on the basis of Article 22 (formerly article 14). This means that as long as a priority of avoidance is not legally embodied by the application of ordinances, decisive action will be based on the consumers' attitude.

The distinction between thermal treatment as an elimination mode and as a means of energy recovery depends on the main purpose for treating waste. Minimum criteria (Article 6 II) facilitate the definition of the main purpose for incineration. If they are met, combustion results in energy recovery in contrast to thermal treatment:

- thermal value of the waste must be at least 11 000 kj/kg;[43]
- a combustion efficiency of at least 75%;
- the resulting heat has to be used;
- further wastes produced during the recovery process can be landfilled without any further treatment.

Instruments. At a secondary legislative level, restrictions may be put into effect within the scope of product stewardship (now Article 22): product stewardship by the manufacturer is designed to achieve prevention at the level of production processes and products. The first instrument referring to such a situation was the Packaging Ordinance. As long as the authorization to define stewardship concerning particular products and organizational needs, is not used, stewardship remains theoretical.

Technical Instruction on Waste from Human Settlements of 1993 (TASi)

Objectives of TASi. The Technical Instruction on Waste from Human Settlements (*Technische Anleitung Siedlungsabfall*, TASi) determines how municipal solid wastes have to be treated by WDAs.[44] This administrative regulation has the objective to minimize the volume of municipal solid wastes to be deposited and to render them inert by turning them into substances similar to ore or the earth's crust, thus preventing contamination through leachate in the future.

Three main treatments are applied to implement the objectives:

- separation of organic waste at source and composting;
- incineration of residual waste before being landfilled;
- deposition on landfills that comply with stringent geological and technical sealing layers.

Regulations have to be fulfilled by the year 2005 allowing 12 years to plan investments.

Investment needs. Investment needs are estimated to be high. This is not only true for the building of incinerators but also for landfill sites that have to conform to different requirements on natural ground and additional systems to tighten up the ground and the surface, also dependent on the content of toxins in landfills.

Although estimates published in 1996 by the Federal Environmental Agency are no longer relevant today, they are still useful to illustrate the main conflicts of implementation.[45]

Composting. The TASi regulates, in particular, the reuse of biodegradable organic substances to reduce waste going to landfills (residual waste). These wastes should be treated with composting methods (with a growing role of fermentation), thus closing product cycles. The reuse of the compost[46] has to comply with minimum quality requirements laid down in the LAGA[47] specification sheet M 10. According to those minimum criteria, the organic material has to be separated at source. Minimum standards also refer to the composting process as well as to facilities requiring capital intensive investments.

The production costs of compost are about 100 ECU/tonne on average, but the retail price is only about 20–40 ECU/tonne.[48] It is estimated that investments for composting facilities will be 1.5 billion ECU for the establishment of 150–160 composting facilities with an average capacity of 30 000 tonnes/year.

After having separated the organic material, residual waste should be pretreated before being landfilled.

Pretreatment of residual waste: incineration or cold pretreatment? Wastes which could not be avoided nor recovered must not contain more than 5% (by weight) of organic components measured in terms of 'volatile solids', which evaporate quickly (*Glühverlust*). In practice, decomposition processes of landfill should no longer occur. The only state-of-the-art technology able to meet the standards on landfills is thermal pretreatment.[49] Whereas the need to pretreat waste before being landfilled is undisputed, the debate focuses on the suitability of the parameter to make landfills safe in future. The interests of the *Länder* differ according to former infrastructural decisions. Those who did not establish a sufficient network of incinerators to match the targets prefer cold pretreatment because investment needs are estimated to be lower. Alliance 90/Greens assume that costs of cold pretreatment (60–75 ECU/tonne) are much lower than the costs for thermal treatment,[50] but this assumption is questionable.

Although the CDU/FDP Federal Government insisted on the criteria of 5%, the SPD opposition and Alliance 90/Greens were in favour of parameters which make cold pretreatment possible. With governmental changes in 1998 it can be assumed that the criteria in favour of incineration will change. Moreover, the *Länder* do not accept that there are in fact no other technical possibilities for pretreatment today.

According to the Federal Environmental Agency, the investment needed within 10 years for incinerators with an average capacity of about 200 000 tonnes/year was estimated at 5.5 billion ECU. The treatment costs would be about 1.5 billion ECU/year.[51] By contrast, another estimation revealed that there is no need for additional incinerators to treat waste from human settlements if there is a recovery rate of at least 60% and a complementary cold pretreatment with a 40% reduction of residual waste.[52] In fact, 13 facilities for thermal treatment are under construction today with a capacity for treatment of about 2.6 million tonnes/year. Costs of thermal treatment will be about 125–250 ECU/tonne.[54]

Landfills. Only two types of landfill will be allowed that comply with the strict requirements for geological and technical barriers. According to the provision of safe landfill sites in future, leachate and escaping gases have to be captured and recovered as far as possible. Landfilling costs are estimated to be 75–200 ECU/m^3, depending on the type of landfill and on the pretreatment applied.[55]

Outcomes of the new approach

Financial burden sharing

Development of license fees in the DSD

The payment of license fees permits a green dot to be printed on packaging. The green dot on sales packaging serves to finance:

- the whole system of collection and sorting;
- the return service for public corporations taking over non-recyclables from sorting;
- public relations;
- administration of the DSD;
- in addition, the costs of plastic packaging recycling.

The criteria for the license fees have changed in the meantime. Until 1.10.1993 the fees were levied per packaging unit only depending on the volume, as is still the case in France for instance. This was the only criterion agreed upon by the packaging industry at the very beginning. The succeeding alteration in the structure of license fees was a result of improved information on material specific costs of recovery:[56] in October 1993, the basis of assessment was altered from volume to weight and material specific fees to cover costs according to the polluter pays principle. The alteration led to increased fees for glass and plastic. The last alteration occurred in 1994 to account for material specific costs and to cover handling costs for sorting, especially those referring to small packaging. Therefore, each packaging unit is now additionally charged with 0.06 ECU on average.[57] Table 5 illustrates the last alteration in the structure of the fees.

Table 5. The amended structure of license fees (ECU/tonne)

	Since 1.10.93	Since 1.10.94
Glass	80	75
Paper	165	200
Tin plate	280	280
Aluminium	500	750
Plastics	1500	1450
Composites		
• Boxes for liquids	830	845
• Others	830	1050
Natural materials	100	100

Source: Clemens (1994), p. 17.

Whereas fillers, producers and importers pay the fee, there is no doubt that consumers have to bear at least part of the financial burden effectively. On average, expenditures per consumer were 19 ECU in 1993, 21 ECU in 1994 and 24.5 ECU in 1995 and the following years.

Due to a degree of licensing of 70% in 1995, returns from licensing were 2.05 billion ECU in 1995. In the succeeding years, the income from license fees could be stabilized at this level. DSD intends, however, to reduce license fees to 1.7 billion ECU until the year 2006.[58] The income serves primarily to finance collection and sorting: the contracts with the disposal branch cause costs of about 1.95 billion ECU annually. About 250 million ECU are used to finance plastic recycling.

Figure 3 shows the financial burden-sharing of the German recovery scheme for respective industries. The financial burden referring to different packaging materials and their weight[59] is illustrated by the order of costs in the beverage market (including 0.0045 ECU per packaging unit):

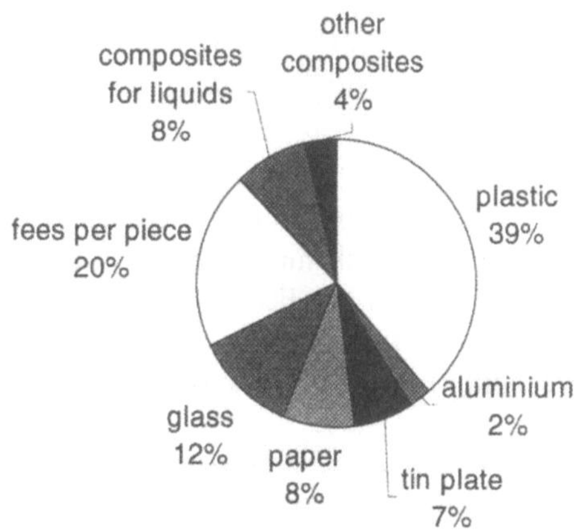

Figure 3. License fees paid for packaging materials (2 billion ECU on a total).
Source: DSD Geschäftsbericht 1997, p. 13

Plastic bottle (80 g weight)	0.1200 ECU
Non-returnable glass bottle (up to 600 g weight)	0.0495 ECU
Composite (26 g weight)	0.0265 ECU
Tin plate can (35 g weight)	0.0143 ECU
Aluminium can (20 g weight)	0.0195 ECU

Increase of costs and fees in public waste management
The relief of landfills due to packaging recovery was estimated at 7% by weight of all wastes per year, which are supplied to household waste landfill sites.[60] WDAs could, *ceteris paribus*, diminish local fees by 250 million ECU/year. However, the potential relief could only compensate for increasing costs. Figure 4 shows the increase of prices on average from 1972 until now and the forecast until 2005 based on the assumption that the TASi requirements are enforced.

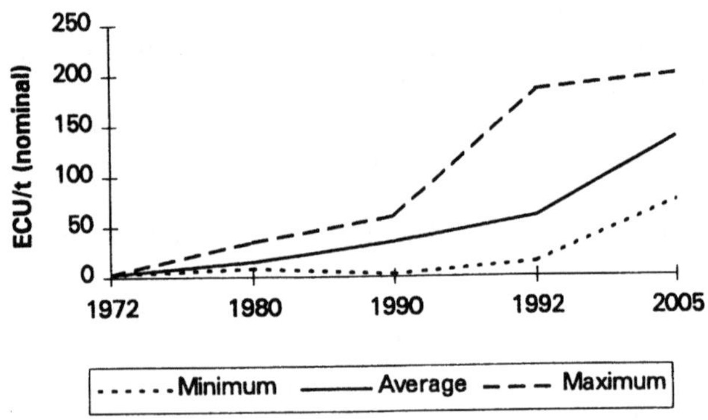

Figure 4. Development of prices for public waste disposal.
Source: SRU (1990), p. 443

The sharp increase in costs at the beginning of the 1990s is mainly due to investments in flue gas cleaning which made up about two-thirds of investments in incinerators. According to the Federal Environmental Agency, the cost increase was not caused mainly by environmental legislation, but by threshold values far exceeding regulations due to the negotiations with local NGOs.[61]

There are several reasons for the *waste emergency* now occurring such as growing efforts in source reduction, new forms of recovery which have developed such as the filling of old mines, the possibility of energy recovery in industrial facilities such as cement factories, and the possible export of waste for recovery to countries with lower environmental standards.

Strengthened by the inherent character of sunk-costs of waste treatment facilities, operators of landfill sites are trying to acquire as much waste as possible before the TASi requirements come into force. Changing market conditions due to the radical change from a disposal emergency into a waste emergency and the current decline of landfilling prices can also be explained by the underuse of public facilities.[62] Industry and commerce are obviously looking for cheaper options for their waste in the market-oriented regime in case of an intended recovery.

Since industrial waste which is delivered to public facilities shrank in a radical way in some regions, average costs are rising. The instrument of compulsory usage helps to cover increased costs by raising local fees: in contrast to industrial waste producers, private households cannot escape either compulsory usage of waste bins or compulsory usage of facilities. They have to bear the financial burden of sunken costs instead of benefiting from the deregulation and decreasing prices. The contrary is true for industrial waste for recovery: lucrative markets which respond to price signals are developing quickly for this waste. Consequently, there is pressure to lower the cost of waste treatment. The interaction between public and private waste management regimes illustrates that the development of the household waste regime and especially the problem of capacity use is closely connected to the waste regime of industry.

The increase in average costs[63] has in fact consequences for households. Between the mid-1980s and 1990s fees have more than quadrupled.[64] Figure 5 illustrates some main features of the development of costs in MSW management.

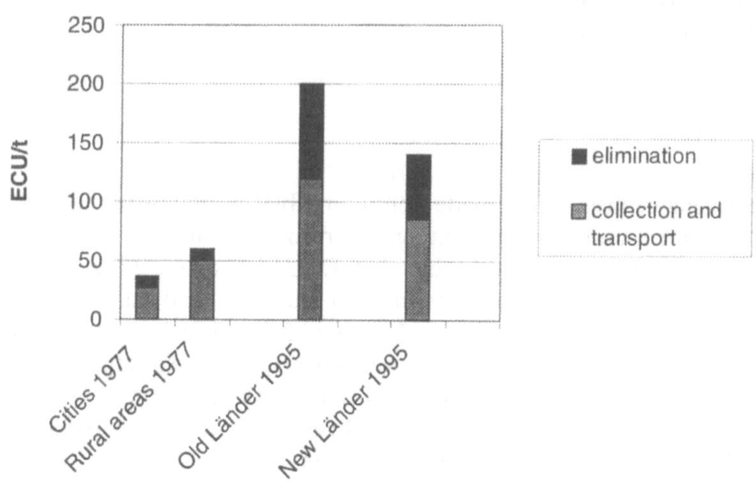

Figure 5. The difference of costs – 4 examples.
Source: Doedens (1982) and Behrens and Maydell (1998)

Fees are higher if previous infrastructural decisions were in favour of incineration. WDAs which decided to introduce extra bins to separate organic waste at source also had to raise fees considerably. The example of *Bodensee-kreis* (no incineration) illustrates that the percentage of costs for composting is 43% of the total fees, whereas only 31% is used to finance the collection of residual waste, two transfer stations and one landfill site.

The structure of costs in MSW management can be summarized as follows:[65]

- Public waste management costs on average: 70 ECU/inhabitant annually, ranging, however, between 20 and 135 ECU/inhabitant, or 176 ECU/tonne on average;
- Costs of thermal treatment: 185 ECU/tonne on average, ranging between 62 and 425 ECU/tonne;
- Costs of landfilling: 75 ECU/tonne on average, ranging between 5 and 225 ECU/tonne;
- Costs of composting: 200 ECU/tonne on average, ranging between 45 and 500 ECU/tonne;
- There are quite significant differences between the old and the new *Länder* where especially landfilling costs are lower (only a few landfill sites comply with the state-of-the-art standards);
- Costs ranges in the old *Länder* vary widely depending on pretreatment applied and the accomplishment of TASi requirements;
- Costs are not considerably influenced by the organization of MSW management: there is only a tendency for lower costs with private waste management companies.

Effects on waste generation

The system of refillables serves as a trigger

The raising of license fees made non-returnables more costly than refillables: the costs for producing, distributing, and recovery of non-returnables per litre beer exceed those of refillables by 0.065 ECU.[66] The percentage of refillables decreased by 1% annually between 1970 and 1985 and has been stabilized today at a level of 72% (in the new *Länder* 58.5% and 75.1% in the old *Länder*; with respect to milk: 22.5%), the level targeted in the Packaging Ordinance. However, the system was questioned when eco-balances were carried out in the last years. The advantage of source reduction has to be compared with the ecological disadvantage of an increase in traffic caused by transport of refillables.

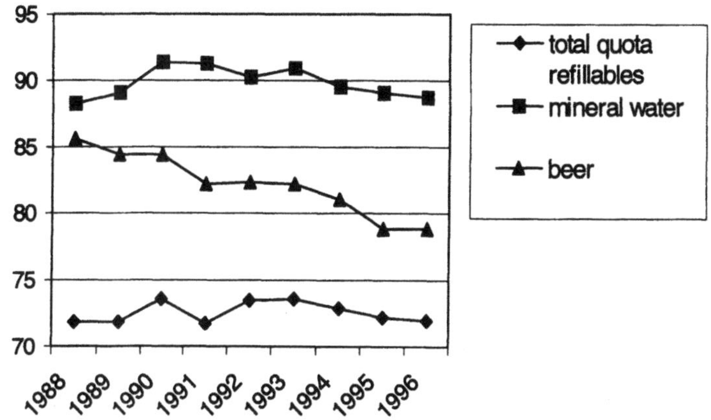

Figure 6. Examples for the development of realized quotas concerning refillables. Source: GVM (1998), press release

The system of refillables can only be realized in the glass and plastics industries. The glass industry has the highest proportion in total packaging of beverages at 75%. The three examples in Figure 6 illustrate that there is a movement in packaging markets, mainly dependent on possibilities for substitution in filling and distributing processes[67] and not least on the acceptance of materials by consumers:

- The sharp decrease for beer means that glass is being increasingly replaced by metal cans (mostly tin plate). Non-returnables are used for market-penetration in the new *Länder* as well as abroad, mainly by big German fillers.
- With the slight decrease in demand for mineral water it can be assumed that glass is being replaced by PET. Starting at a low level, there is an increasing percentage of returnables which will not be re-used,[68] but which are said to be recycled. French and Belgian fillers of mineral water make particular use of deposit–refund systems: PET bottles are charged with 0.25 ECU deposits. Moreover, the German filling association GDB announced in 1996 a switch from the standardized type of bottle (*Perlglasflasche*) to refillable PET bottles because of international pressure: the radius for action should be enlarged by the lighter packaging material, reducing transportation costs by 40%. Thus, plastic increasingly replaces glass as traditional packaging material on the mineral water market.
- The German quota for refillables is still quite high. One important reason is the behaviour of consumers who buy beverages mostly in special beverage markets that supply refillables. The food retailers only play a minor role in

this respect. For mineral water and some specialized beer markets (*Bavarian Weizen*) the means of standardization as a voluntary co-operative action of fillers to reduce transaction costs made the system of refillables economically efficient.

The economic advantage of refillables, however finds its counterpart in the potential returns from non-returnables. There are several reasons for their attractiveness, thus promising higher returns which compensate higher costs.

Enacting mandatory deposit–refund systems

It was foreseen in the Packaging Ordinance that if the beverage industry did not achieve the targeted 72% quota, non-refillables would be excluded from the licensing system DSD. The effectiveness of mandatory deposit–refund systems to serve as a deterrent from not achieving the quota of 72%, is mainly dependent on the expected additional costs of organizing deposit–refund systems. A study[69] revealed that costs of deposit–refund systems are comparable to the license fees paid, due to the possibility of providing automatic deposit containers near points of sale. The system could be financed by those consumers who do not bring back charged bottles; the break-even point is at a maximum circulation of 90%. In the case of plastic bottles, the costs of deposit–refund systems (non-refillables) seem to be even lower: Belgian and French fillers of mineral water started to charge (PET) bottles with deposits of 0.25 ECU in order to escape the German recovery system. It is said that the trigger does not function because in the end, the main objective is not achieved, i.e. the stabilization of the system of refillables.

The deterrent, however, if applied, would cause a loss of DSD's income from license fees of 25% (500 million ECU) – corresponding to 475 000 tonnes which would have to be collected by retailers. Assuming the payments of DSD to collectors to be proportional to their achievements, 25% of the market would break down. DSD will have to try hard in this case to reach the defined recycling quotas for the first time because the high quotas reached in the DSD are said to be mainly due to the high circulation of non-refillables. The logic of the Packaging Ordinance would lead to the abandoning of the system.

The development of packaging use

Packaging use has decreased in total since 1992. Sales packaging makes up 55–60% of total packaging. Since the enacting of the Packaging Ordinance, reduction of weight is significantly important in the field of secondary packaging (decreasing by nearly 50%), only affecting the paper and plastics industries.[70] Moreover, the trend towards lean packaging further reduced the weight of packaging, a trend which had started in the 1970s to save costs.[71] Despite the 12% reduction of sales packaging use between 1991 and 1995, at least one-third of this tendency was due to the general decrease in demand,[72] the most recent tendency shows an increase. After a slight increase of sales packaging use in 1996,[73] there was a considerable increase in 1997 (Figure 7).

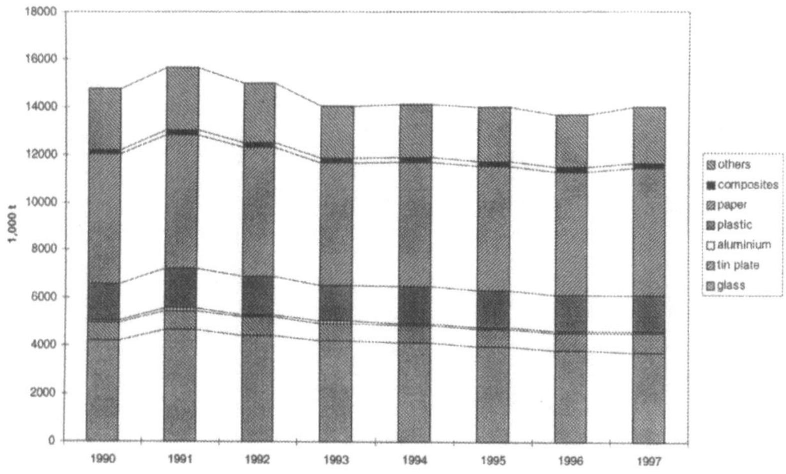

Figure 7. The development of total packaging use.
Source: GVM (1998): press release

Recycling achievements

Not least due to the existing infrastructures for the collection of glass and paper, the DSD was able to collect more than 4.6 million tonnes (57% of used packaging) in its first year (Figure 8). Excluding aluminium and plastic in the first years only, the realized recycling quotas were generally higher than those prescribed by the Packaging Ordinance. Figure 9 illustrates the costs caused by the increased recycling level.

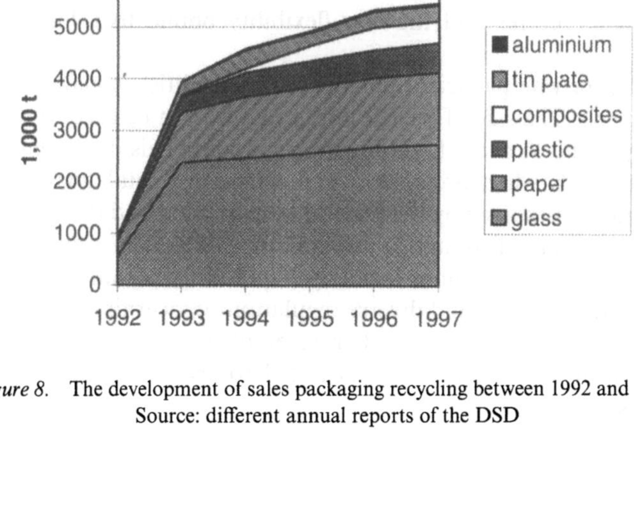

Figure 8. The development of sales packaging recycling between 1992 and 1997.
Source: different annual reports of the DSD

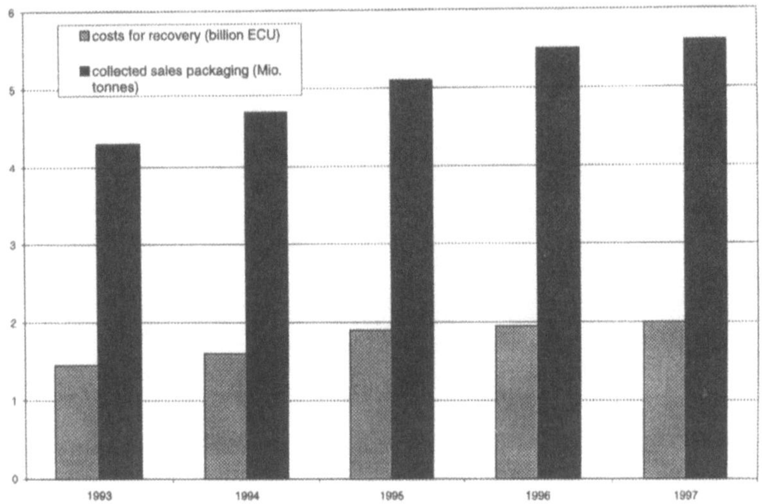

Figure 9. Recovery costs of increased recycling achievements.
Source: annual reports of DSD

The organization of waste treatment in the 1990s

Tendencies towards privatization and concentration. The development of public private partnerships. In the 1990s, especially in some of the cities, public firms were quasi-privatized (*Eigengesellschaften* mostly run as plcs).[74] The trend towards public private partnership concerning waste treatment activities has been promoted as a future model by the Association of Municipal Waste Economics because it would benefit from the advantages of both organizational forms (efficiency and private capital combined with sovereign tasks). This form, however, was criticized due to private capital profits from the monopolistic structures and the new flexibility bound to a loss of political control.

The current private participation concerning facilities is mainly focused on financial aspects. The sale of partial infrastructure can be used to stabilize the public budgets and to avoid high investments today. This development started in the early 1990s when ordinances and normative catalogues of technical solutions had been enacted by the Federal Government and the *Länder*. Due to ever increasing standards and above all the TASi demands, much larger investments than the minimum formerly required for a sanitary landfill were now necessary. The many and diverse regulations concerning the development of the infrastructure, the expected transfer of expertise and primarily, the limited financial capacity of the WDAs led to the described public–private partnerships.

The trend towards privatization conflicts with capacity use by WDAs. The increase in the number of employees in the 1980s and 1990s indicates that the private sector became the driving force in the dynamics of additional employment (Table 6).

Table 6. Development of the number of employees in the years 1983 to 1993

	Employees in private enterprises	Employees in public enterprises
1983	16942	48528
1989	24029	51932
1993	45238	56907

Source: Bundestagsdrucksache 12/ 6685

This was especially true in the new *Länder* after 1989 (the year of German unification) where there had been no tradition of self-regulating companies. Waste management tasks such as transport and collection were carried out by private firms, because of the difficulties involved in setting up administrative structures. Not surprisingly, turnovers within the sector also increased,[75] nearly quadrupling between 1980 and 1990 (nominal).

A substantial change leading to privatization was constituted by the Closed Substance Cycle Waste Management Act 1996 because it was then no longer obligatory to make wastes which had originated in commercial and industrial sectors available for treatment to WDAs and public facilities (Article 13). In the case of domestic waste, however, households were still obliged to use bins and public facilities, except for organic waste which could be composted.

Due to decreasing waste amounts today and therefore increasingly underused waste treatment facilities after the recent development of a very complex infrastructure, WDAs are, however, interested in the possible application of instruments such as the obligation to make wastes available to them and the compulsory usage of specified plants for further treatment.

The concentration process in waste management. The main driving force towards concentration comprises benefits resulting from economies of scale. The number of mergers rose from 30 in 1989 to more than 120 in 1993.[76] A number of medium-sized companies have been taken over by some of the newly formed giants in waste management such as RWE and VEW, both energy sector companies.[77] Furthermore, there is a lack of demarcation between public administration and privately operated enterprises.

The waste business is of special interest for the energy sector. Energy companies are looking for fields of diversification, primarily because of the liberalization of the energy market in Europe. The crucial point is that there has been a long tradition of regulated co-operation between public administration and the coal, energy, and steel industries. Until recently protected, the energy sector is now looking for those fields of diversification which are just as strongly regulated.

The small and medium-sized waste companies (represented by the association BVSE) in particular, as well as the Federal Cartel Agency, pointed out that the environmental regulation setting high standards and high targets tends to have rather problematic effects on competition caused by the corporate behaviour in this sector in order to overcome fee avoiders exploiting the system (free-rider positions) and to reach targets in a cost efficient way. When the DSD was established by industry and retailers, the assumed tendencies towards monopolization were tolerated by the Federal Cartel Agency in order to give industry a chance to meet targets set by the Packaging Ordinance.

An example of the lack of competition is the guarantor for plastic packaging recycling, the DKR.[78] Partners are DSD (25%), plastic producers and manufacturers (25%), and the disposal branch with 50%. This partnership led to the lack of competition in plastic recycling because disposal enterprises could determine prices as well as quantities (demand for the collected substances *and* supply of capacities). The possibility for competitors to enter the market was limited because there were 10-year contracts between the DKR and disposal enterprises. With the interdiction of the Federal Cartel Office in June 1994, those contracts were dissolved.[79]

Environmental legislation and in particular the instruments of command and control show that monopolization goes hand in hand with the potential to benefit from economies of scale. Due to increasing returns of scale, one collector can supply at the lowest average costs, thus creating an inherent trend towards concentration. Lack of competition, however, leads to relatively higher prices required by the monopolistic supplier. A condition to restrict monopolistic behaviour could be 'Demsetz-competition', i.e. independent disposal enterprises from the DSD compete from time to time with each other in order to win contracts. The pressure on the DSD (short time span to establish the system) was misused by disposal enterprises through monopolistic price fixing, also blamed for the financial crisis in 1993.

Development of treatment capacities

Up to 1990, the trend towards centralization had continued, illustrated by the further decrease of landfill sites on the one hand and an increase of transfer stations on the other. In 1990, there were only 310 landfill sites left in the Western part of Germany. However, the number of landfill sites increased to 6300 as a result of German Unification in 1990. Those landfill sites were mainly uncontrolled. According to the Federal Statistics Agency,[80] German landfill sites could have been reduced to a total of 562 in 1993.

Incineration reached 25% of total waste treatment in the old *Länder*. Due to the unification it decreased to 17%. It should be pointed out that there are rather different developments concerning incineration in the different *Länder*, as in the case of Bavaria.

The development of composting facilities especially in the late 1980s was due to the efforts made by WDAs. In 1990, 218 facilities had already been established in the Western part, whereas there were only 12 in the Eastern part. The capacity was 4.1 million tonnes in 1995 due to a further increase to 386 composting facilities in total.[81]

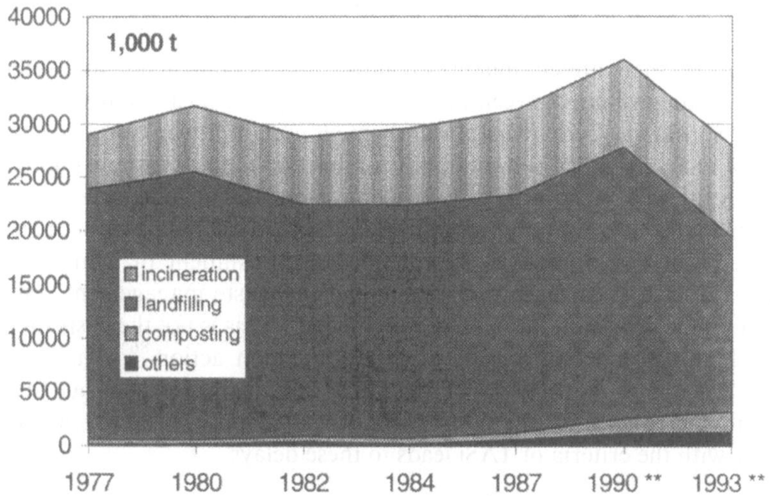

Figure 10. Destinations of waste from human settlements delivered to public facilities.
**Including the new *Länder*.
Source: Statistisches Bundesamt (1996)

Following the enactment of the Packaging Ordinance, a recycling infrastructure developed rapidly. In 1995, 455 sorting facilities and 47 processing facilities (26 for glass and 21 for plastic)[82] had been built. Recycling capacities are said to match the amounts collected today.

Delay in investment in future capacities

First of all, planning procedures have changed: to avoid blocked decisions, local mediation procedures (co-operation of all involved participants) and intensive public relations should create a consensus for waste treatment. Since 1993 the approval of waste treatment facilities has been assisted by the Investment Improvement Act due to the threatening disposal emergency: there is no longer the need to prove the objective necessity of a plant. The criteria for approval are based on the general requirements for industrial plants under the legal regime of the Federal Pollution Control Act. Only landfill sites remain under the special approval procedure of the waste regime.

Once again, as at the beginning of the 1970s, the planning framework has become too small for efficient solutions. The *Länder* responsible for waste disposal plans have begun to encourage co-operation between WDAs.[83]

Uncertainty concerning standards, prices and the surplus or scarcity of treatment capacity still lead to delays in complying with the requirements of TASi: this is especially true for the necessary incineration of household waste and for compulsory composting. There is also some opposition to incineration requirements set at federal level, and the political opposition which won federal elections in 1998 criticized the emission of dioxin as well as the additional incineration capacities which do not comply with the objective to avoid waste.[84]

Moreover, there are complaints about the loss of control due to insufficient criteria distinguishing between waste for recovery and waste for disposal in the Closed Substance Cycle Waste Management Act of 1996. Consequently, all kinds of recovery processes are now available and their environmental impact is often criticized. According to those complaints, the ongoing privatization in waste management not only jeopardizes investments in public waste treatment facilities, but also threatens the intended establishment of a high quality infrastructure to treat hazardous waste and the waste management task as a part of public provision for basic requirements.[85] This is not the basis for long-term planning, and only results in delays and non action.[86] The on-going discussion about the criteria of the TASi and instances such as the first exceptional approval for the WDA Hannover to dispose waste which does not comply with the criteria of TASi leads to these delays.

Correcting evolution in managing packaging waste

Movements of dissatisfaction

Incentives to export and financial supports cause distortions abroad

The strong incentive to avoid domestic costs led to considerable exports to European as well as to non-European countries. In the importing country the 'secondary' commodity could disappear as waste in landfill sites or incinerators due to lack of international control.[87] Although the DSD is obliged to inform the Environmental Ministry of the whereabouts of their collected materials, total control of final destinations is questionable. Complaints addressed to the European Commission forced the DSD to recycle packaging waste in Germany. It was perceived that Europe had to suffer because of the shipments of German packaging waste as a result of the high recycling targets. Consequently, the European Packaging Directive institutionalized the strict self-sufficiency principle for recycling packaging waste in case of national opt-ups, i.e. higher targets and an extended responsibility for producers. However, shipments of packaging waste have become a problem of transboundary trade with secondary materials today. In the case of plastic recycling in particular, markets for the re-granulate (secondary) are characterized by over-capacities and low prices on primary markets.[88] However, the secondary material became attractive due to the high financial support in Germany for collection, sorting and recycling (1450 ECU/

tonne). Distortions occur on secondary markets abroad because the German system on the whole supports collection and sorting by the strict implementation of an extended packaging stewardship, in contrast to other European countries where the shared responsibility principle has been institutionalized (between industry and WDAs). Finally, the problem caused by a lack of facilities for recycling sorted materials has not been solved but rather shifted to other countries, because the materials can be exported at high profits and due to the lack of clear-cut borders between packaging waste and secondary materials which can circulate freely[89] in Europe. As a result of highly subsidized German secondary materials arriving on European markets, traditional collection schemes in neighbouring countries have collapsed because of less financial support. The difference between incomes from license fees in France (73 million ECU/year) and Germany (2 billion ECU/year) clearly demonstrate the difference in financial support for collection and sorting.

European pressure to change the legal frame

The legal situation changed by the European Union was one of the main driving forces for the amendment of the Packaging Ordinance. A letter from the European Commission dated 12th December 1995 addressed to the Environmental Ministry refers to the incompatibility of the existing Packaging Ordinance with the regulations on free trade. The complaint focuses on the quota for refillables which causes the enforcement of mandatory deposits on non-refillables which is interpreted as an obstacle to free trade. Furthermore the European Directive on packaging only allows the setting of higher standards in case of compliance with the self-sufficiency principle. Moreover, participating in the system without paying the license fees should be sanctioned.

The financial crisis of the DSD in 1993

Fee evaders do not pay for the disposal system although they profit from it because their packaging waste is treated in the privately operated system.[90] One of the main reasons for the financial crisis of the DSD in 1993 had been the inadmissible use of the 'green dot': although 90% of all products carried this symbol, only 50% of fillers/producers paid their fees correctly. The financial loss by this fee avoidance was estimated at 400 million ECU per year.[91] In order to minimize fee avoidance, or more precisely to secure the system, retailers supported the DSD by collecting fees (*Handelsinkasso*).

The high fees for collection and sorting agreed upon between the DSD and the disposal branch with the 'service contract zero' were another important reason for the financial crisis. Due to the pressure to establish a recovery scheme covering 100% of households in a short time span, set by the Packaging Ordinance, the disposal branch could successfully enforce high fees in return for their services.[92] The financial crisis in 1993, enabled disposal enterprises to infiltrate the board of directors by transferring part of the debts of the DSD into partnerships.[93]

The power of the retail sector to exercise pressure by announcing they would no longer sell products not labelled with the green dot has been used to generate a stable and goal-oriented co-operation between market partners involved. The stability of this sanctioning mechanism was threatened, however, by the incentive of fee avoidance. Single retailers selling their own brand goods have ignored the agreement to save on license fees: their brand goods do not carry the green dot although most of their packaging waste ends up in the yellow bin provided by the green dot recovery scheme: they are not obliged to take back packaging waste individually.[94] The sooner the liable group seems to achieve the required quotas, the greater the incentive is for individual companies to behave unco-operatively.[95] With respect to internal stability, the comparative high quota can be seen as a factor limiting the exploitation of fee avoidance. The amendment of the Packaging Ordinance in 1998, however, only succeeded because big retailers (e.g. Metro AG) announced their intention to leave the DSD because of ongoing free-riding and the lack of a legal deterrent which could be applied. Their announcement forced the *Länder* to agree with the amendment because otherwise the DSD would have collapsed due to a threatened break-down of income from license fees.

The debate about an amendment of the Packaging Ordinance reveals economic interests

Failures of command and control are due to the State not having sufficient information to determine efficient quotas, and coupled with this, influence is exerted on the legislator from interest groups to enforce distributional interests. The owners of those interests will therefore bargain hard to protect their investments. At the hearing of parties concerned with 140 associations and interest groups, the Association of German Industry (BDI) agreed with the European Commission's point of view that interventions such as quotas for refillables are illegal beyond the development of the single European market.

The Industrial Association on Plastic Packaging (IK) criticized the strong interventionism in German waste management in contrast to the European Directive. This association intended to leave the quota of plastic recycling to markets. The same criticism came from the Association of European plastics treatment (eupc): a determination of quotas would neither comply with the Directive nor be technically feasible.

On the contrary, the Federal Association Secondary Raw Material and Disposal (bsve) representing 500 small and medium enterprises for disposal, called for an increase in plastic packaging recycling to 80%, stressing that the potential of material recycling has not yet been exhausted. Moreover, they feared competition would be endangered due to the strengthened position of the DSD. The Association of the German Waste Economy (BDE) also saw their high investments in the establishment of a private disposal system endangered because of an intended reduction of recycling quotas and the change of the basis of proportioning (estimated effect: 2 million tonnes less packaging waste reaching the recycling loop).

Perhaps the most important development in the debate is demonstrated by the *Länder* because of their influence on legislation: some of the *Länder* have demonstrated their interest in combustible waste because their incineration facilities are underused and they are forced to substitute waste for fuel.[96] This situation led to the political demand that they should be allowed to incinerate small plastic packaging (constituting 61% of plastic waste from the DSD collection). Plastic recycling will remain one of the controversial issues in MSW management in future.

Towards an equilibrium

The amendment of the Packaging Ordinance in 1998

The most important amendments refer to the development of a new basis for the assessment of recycling quotas. Instead of total packaging use, the new basis refers to the licensed amount of packaging. Moreover, if not participating in the privately operated recovery scheme, compliance with the requirements has to be proved individually. Although the quotas could be reached already, targets have been lowered. With respect to plastic packaging recycling, quotas have been lowered and split: 60% has to be recovered in total, and at least 36% has to be recycled. The current trend towards (cheaper) feedstock recycling such as the 'gasification' in steel mills as a substitute for fuel is now limited to a maximum of 24%.

With respect to the quota on refillables as a trigger to enact mandatory deposits on non-refillables, the amendment of the ordinance weakened the possible application of the deterrent: if the quota is not met, at least 2 years is allowed until a renewed control of the quota. In the meantime, the European Court of Justice will probably judge on the legitimacy of the German approach.

The amendments of contracts between the DSD and the disposal branch

With the 'service contract zero' in 1991, the following payments had been agreed upon:

- 75 ECU/tonne for the collection of glass;
- 100 ECU/tonne for collection and sorting of paper;
- 750–900 ECU/tonne on average for collection and sorting of metals, plastic and composites.

In contrast to supports of Eco-Emballages, German disposal enterprises had been paid according to the input for sorting facilities on average per tonne. As a return service for the high payments, they had to leave the sorted materials free of charge to guarantors, known as the 'intersection point zero'.

With the first and second amendment after the financial crisis in 1993, the final payments were limited to 20 ECU per inhabitant on average in 1994 which is, in fact, a 'lump sum revenue'. The third amendment aimed at a higher quality of sorting: instead of inputs, they are paid according to outputs of sorting and financial supports are progressively increasing thus promising more cost-efficiency (up to 12 categories). Moreover, there is the possibility with the new contracts to transfer collected and sorted quantities which exceed the required quotas to another disposal enterprise in order to reduce the inefficiency of covering 100% of households. This (*bubble policy*) approach, however, is only applicable within one *Land.* To meet the demands of the European Commission, the 'intersection point zero' had to be dropped because of the distortions caused on secondary markets. With the third amendment, sorters are free to sell the sorted materials now. However, in practice sorters still leave the secondary materials free of charge to guarantors because they have to take over the risk and have to pay 0.6 ECU/tonne for administrative costs to the guarantor.

The third amendment reflects the increasing importance of cost-efficiency. This is of course an outcome of the international pressure: as neighbouring countries reject a comparable expensive system, German industry is in a bad position. On the other hand, the interests of the disposal branch who benefited from growth rates of about 20% even in times of an overall recession, are well represented in the system which is a German particularity.

The dynamic of plastic processing in Germany

The pressure from the retail sector and producers with the help of the green dot being used as an insider–outsider label ignored the interests of the chemical industry. Due to public pressure, channelled by the most important industrial associations, the guarantor VGK had been founded by the Industrial Association for Packaging and Foils (IK) ensuring the recycling of collected plastic in order to acquire the licenses for the green dot. Whereas existing recycling capacities for the chemical industry could only match 50 000 tonnes/year, the amounts of plastic packaging collected had been estimated at 80 000 tonnes in 1992, 160 000 tonnes in 1993 and 400 000–600 000 tonnes in 1996.[97] Despite insufficient capacities, recycling guarantees were given in order to avoid further legislative constraints and to ensure their participation in the DSD. After the VGK followed illegal paths of disposal such as dumping and incineration abroad, the DSD initiated and participated in the foundation of the DKR as the new guarantor to secure the system.

Although material recycling was required by the Packaging Ordinance, there was a trend for comparison of costs with alternatives after the DSD and DKR were forced into a self-commitment of self-sufficiency due to international pressure. Among the emerging technology options, feedstock recycling became the most important, because of the high proportion of dirty, mixed and flexible packaging (61% within the plastic fraction). With the exception of gasification in steel mills, however, options like pyrolysis, hydrogenation and chemolysis processes are very expensive. Steel mills are in the best position because investments and average costs are rather low. However, the plastic waste has to be processed before being used in steel mills. Due to the savings through input substitution and the additional financial supports of 100 ECU/tonne plastic packaging waste recovered, the *Bremer Klöckner Werke* (steel works) was saved from closure.[98] In this context, there was a good chance for the DSD to develop a processing technology for this flexible fraction in co-operation with chemical industry and a recycler applying this technological innovation.[99] The developed procedure (*dry agglomeration*) needing less energy to process plastic waste is probably only interesting with respect to forthcoming ordinances, e.g. on electronic items, and with respect to technology transfers, managed by SYSTEC, a subsidiary company of the DSD. The patent for the gasification of processed plastic as a substitute for fuel in steel mills is also owned by the SYSTEC.

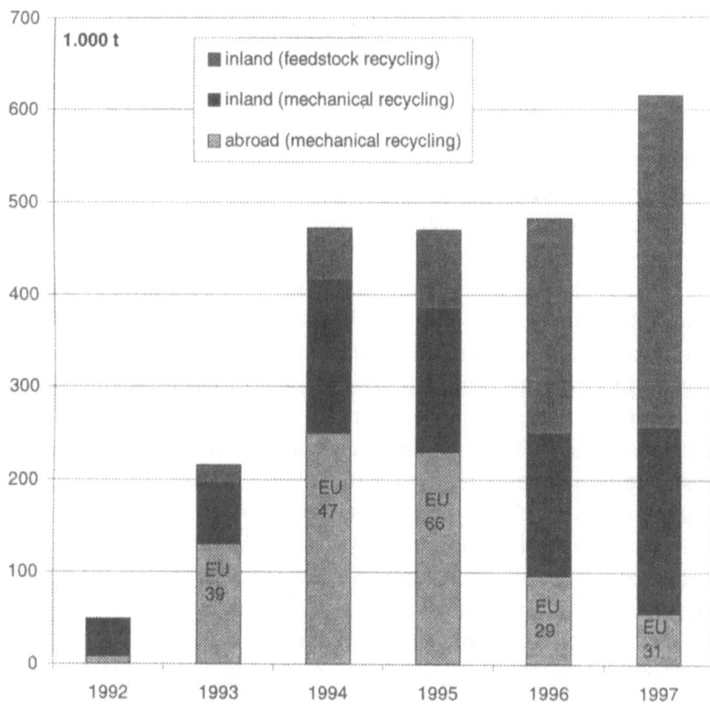

Figure 11. The development of plastic recycling between 1992 and 1997.
Source: DKR (1998): Auf einen Blick

Figure 11 shows that the high recycling achievements in 1993 were mainly due to recycling abroad. The greatest part should have been recycled in China. However, with the resistance of the Chinese population, imports of German plastic waste had to be stopped in 1995. Moreover, the Council and the Parliament voted for self-sufficiency in case of national higher standards. The figure also illustrates that feedstock recycling started to increase in 1994 with the use of plastic waste in steel mills. In 1997 the proportion of feedstock recycling (57%) exceeded material recycling (43%) for the first time. The DKR intends to stabilize exports of plastic packaging waste at a level of 10% and to reduce financial supports for the recycling to zero in the year 2003. There is an evident tendency towards more cost-efficiency in managing packaging waste, requiring cheap solutions such as their gasification in steel mills. The problem of mixed, dirty and flexible plastic (which accounts for 60% within this material group) determines the high proportion of feedstock recycling applied. The amendment of 1998, however, tries to stop these market developments by determining the proportion between material and feedstock recycling to reach the new and lowered recovery quota of 60%: material recycling should make up at least 60%. Thus, the amendment distinguishes between a quota of 24% for feedstock recycling and a material recycling quota of 36%. The undermining of the ambitious idea to close substance cycles aims at an improved cost-efficiency.

Source reduction being the first priority

The increase in waste volumes after the introduction of recovery schemes was observed after collection containers had been established by WDAs for the recycling of non-refillable glass. From the consumer's point of view non-returnable packaging was no longer linked with environmental damage.[100] The German experience demonstrates that efforts towards recycling, interpreted as an enlargement of former constraints, had always been followed by a sharp increase in demand. The criticism of the environmental association BUND is directed at the ploy of the green dot. Intended or not, the consumer is deceived because the colour green is associated with environmental issues and it is assumed that separating (non-returnable) packaging from other waste is an adequate means to relieve the environment. The positive ecological image nonetheless helped to realize unexpectedly high collection quotas.

The establishment of recycling capacities is criticized by the environmental movement, thus preventing efforts of waste avoidance because of their sunk cost character. Several thousands of employees are a factual constraint. In addition, it has not yet been defined in which way the proportion of 72% refillables has to be maintained to serve as a trigger to enforce mandatory deposits. The problem of the trigger is that single fillers do not have any room to manoeuvre: the quota constitutes a community which is in sum liable, but which can neither control nor be responsible for the proportions.

The resistance of German industry against further interventions in the beverage market could support the European Commission's point of view that the regulation serves as a non-tariff barrier to trade. They even requested that the German regulation will be prohibited.

Conclusion: cost-effectiveness becomes the driving force towards an equilibrium in packaging waste management

There are several reasons for the high costs of the recovery scheme applied in Germany:

- The high targets in order to implement the idea of closed substance cycles increased costs on average considerably;
- With the ban on thermal recovery for packaging waste, especially costs for plastic recycling, exploded, those are included in the license fees;
- License fees are increased according to a 'total cost basis' to cover collection and sorting costs in contrast to the 'shared-responsibility principle' in other Member States where only additional costs are covered by financial supports;
- The covering of all households increased costs;
- The high failure quotas as a result of throwing non-recyclables into the yellow bins provided by the green dot recovery scheme: WDAs could enforce high payments for the return service of taking over non-recyclables from the DSD (about 175 ECU/tonne).
- Neither the disposal branch nor the DSD have an incentive to reduce costs. The recovery system is financed by license fees which could be interpreted as an indirect private tax levied on nearly every transaction between retailers and consumers. The effective burden is said to lie mainly with consumers (25 ECU per inhabitant annually). The distribution of this tax income takes place within the disposal branch. There are no longer clear-cut borders between DSD and contractors such as the Interseroh S.A. Coincidentally, the enterprise had been founded in 1991 nearby DSD at Cologne-Porz.[101] Except for glass, Interseroh S.A. is guarantor for each material; 80% of all contracts in the DSD are associated with Interseroh. The external growth of the enterprise has to be financed by consumers and partially by importers, subsidizing the development of the lead position of Germany on worldwide secondary markets. DSD, transferred into a AG in 1997, itself is subsidized by the tax income, thus being first on the technology market. The two patents promise high profits from selling the licenses for the processing of plastic (agglomeration) and the succeeding use in steel mills.

Due to the fact that the European Directive on Packaging and Packaging Waste remained far behind the German requirements, the pressure to reduce costs increased further. Due to steadily increasing and stable waste flows for the DSD, investments were made. The focus is on automatic sorting with an integrated processing system which will be presented on a large scale at the EXPO 2000 for the first time. The amendments of contracts between DSD and the disposal branch succeeded in improving sorting qualities and in stabilizing costs although sorting achievements are still increasing.

The intended reduction of license fees from 2 billion ECU to 1.7 billion ECU in the year 2006, will be mainly reached by cost reductions in sorting (estimated at 50%) and plastic processing and recycling: financial supports for the recycling of plastic packaging at the current level of 250 million ECU are intended to be reduced to zero in the year 2003. As the DSD tries to levy license fees according to specific recovery costs, it can be assumed that the plastic industry will benefit most from the realized cost reduction.

One district in Hessen intended to cancel contracts with the DSD and made contracts with the inventor of a dry agglomeration process which reduced the volume of waste by 40% due to the extraction of humidity. The new contractor promises to reduce collection and transportation costs by 50% by simplifying the collection system: only paper and composites with a paper basis will be separated at source, glass and hollow plastic packaging are collected in containers near points of sale. The other materials will be sorted from the residual waste by eddy current separators and magnetic separation. The separated paper will be directly transported to paper mills. The residue from the dry agglomeration process will be incinerated in industrial facilities due to the high heating value of the product, comparable to brown coal. Although emissions are said to be higher than in the case of feedstock recycling, the WDAs and the Environmental Ministry are showing a keen interest in the pilot project.

Experiences and prospects

Changes in the orientation of the German waste disposal regime, from a public provision for basic requirements towards a strongly market-oriented policy, have been described in this chapter. The development from a public health regime towards a waste management regime that was still, however, marked by traditional ideas required the further development of existing institutions and the creation of new institutions. This was necessary to accomplish the *Hochzonung*, i.e. the transfer of power from the local to the regional level and/or federal level which made the already progressive procedure of waste treatment economically viable.

German waste management policy relies basically on the agenda-setting power of environmental programmes and experts' discussions. In practice, however, the demands although translated into political objectives, could not be enforced effectively. This is clearly the case in the packaging sector. The unsuccessful commitments of industry to stabilize the percentage of refillables, and not least, the handing over of the newly institutionalized Ministry of Environment to Mr. Töpfer brought about a change in regulatory style, moving from co-operation (or more precisely corporatism) towards interventions. The awareness of declining future capacities (known as the disposal emergency) and the co-operative demands of non-central political institutions put pressure on the regime to move effectively towards a new regime: the Packaging

Ordinance as a far-reaching instrument for privatization is an illustrative example of the new approach to German municipal solid waste management after nearly 20 years of discussions on the political agenda, which started with the Waste Management Programme in 1975. The extent to which industry could enforce their interests in the bargaining processes is restricted by the strong pressure of the German environmental movement which led to the described higher targets and standards compared to other countries. The Packaging Ordinance reflects the stable focus of discussions on beverage packaging and the strong opposition against incineration, resulting in a ban on thermal treatment of packaging waste. The *Länder* with their influence on legislation also succeeded in introducing compulsory material recycling and the high quota for refillables.

There are two outstanding differences in the new approach to management: increasingly more responsibility is given to the private companies in the disposal sector, and an extended responsibility is taken on by products. Up to now, however, a coherent approach for a new regime could only be achieved to a limited extent. At present, the developments are marked by a series of errors and uncertainties, not least due to the call for structural changes with the enacting of the Technical Instruction on Waste from Human Settlements (TASi) and of the Closed Substance Cycle and Waste Management Act (Krw-/AbfG).

As a result of the introduction of the TASi (including minimum requirements for landfill site standards and the pretreatment of waste), competition between the owners of landfill sites increased since the insufficient standards of their sites would mean they would no longer be allowed to be used after the year 2005. Their interests are, therefore, concerned with the filling of their sites in order to at least partially cover their fixed costs. At the same time, even the standards of the TASi were scrutinized when stipulated incineration was criticized by the new government. The debate on compulsory usage of incinerators to meet the standards referring to waste going to landfill sites (TASi), reflects the resistance of WDAs to incineration as the state-of-the-art standard in pretreating waste, required by the federal level.

Parallel to this, the Closed Substance Cycle and Waste Management Act (Krw-/AbfG) set down the following objectives:

- The responsibility of producers should lead to the avoidance of waste at source
- The reuse of waste should have priority over disposal
- Those responsible for waste disposal should be committed to dispose of residual waste in domestic sites with high environmental standards.

The realization of the objectives of Krw-/AbfG, however, primarily depends on the enactment of the necessary regulations (e.g. how can the responsibility of the producers for their products be put in concrete terms), which has not yet been realized. Concrete terms can only be enforced by issuing statutory ordinances or technical instructions, negotiated between the interest groups concerned. Approaches involving far-reaching privatization and product responsibility can only be seen in outline. Moreover, problems of definition and limitations could not be removed by the comprehensive legislation.

There are two main demands resulting from this which lead to substantial uncertainties concerning the future of the existing waste management regime in the following years:

- on the domestic front due to effects of the Krw-/AbfG and the TASi,
- on the European front due to the development of European legislation on waste, but also the question in how far free trade render traditional principles (such as the proximity principle) obsolete.

With the Krw-/AbfG, waste streams will be at least partially removed from the public regime: a lack of differentiation criteria are the main reasons for the continuing *deregulation.* The clarification of criteria for distinguishing between waste for recovery and waste for disposal determines the scope of a private and a public regime: this is certainly the case for the restricted obligation to make waste available to WDAs in the commercial sector.

Since reusable waste substances can be freely traded as economic commodities, and due to the difference in costs throughout Europe, there is an incentive to mark waste as reusable. It is no longer necessary to rely on the cost-intensive disposal infrastructure provided by the state because financially, the cheapest method of disposal can be chosen or contrived; this is, however, only true for reusable substances. The endorsement of energy recovery (under different standards in different facilities) means that cheaper methods of waste reuse will be found and thus the existing public facilities, which have been established to reduce emission, are devaluated as a result of competing with industrial or foreign facilities.

With the turning away from a central public waste disposal regime (public provision for basic requirements) towards a far-reaching privatization in the waste sector, the state increasingly loses control over the waste streams. Since the definition of whether reusing waste is beneficial or not is decisive in whether cheaper paths of disposal can or cannot be used, this will become a loophole in the whole legislative framework which can be exploited in order to avoid stringent standards.

In total, this results in the underemployment of the existing clean infrastructures; a significant reduction in the waste streams of up to 70% of the commercial waste in the landfill sites and public incinerators has been determined. A problematic role is played by the waste which has been declared as reusable waste which is then reused in facilities with significantly lower environmental standards and thus lower costs. Public incinerators and landfill sites are thus increasingly forced to compete with industrial facilities and other emerging recovery options such as the filling of mines. The unfair competition between clean, and therefore, expensive, facilities and fuel substitution in industrial facilities results primarily from the different technical regulations controlling the facilities: waste incineration facilities are subjected to the considerably stricter standards of the 17th Federal Pollution Control Ordinance unlike the industrial facilities which use the waste for energy and are mostly authorized under the lower standards of the Technical Instruction on Air.

These may be problems of the radical changes to structure, e.g. until the final realization of the TASi which will result in the closure of a large number of landfill sites, but at the same time it is clear which innovative forces will come about from the opening of new cheaper methods of using waste: with the easing of the traditional waste disposal regime and combined with the waste business, numerable alternative paths for the management of waste will be discovered which are cheaper than the traditional methods. The similarity between secondaries and market goods, however, means that a possible re-regulation of secondary material markets at a national level highly improbable in view of the development of common markets and especially the awaited judgement of the European Court of Justice.

Since the privatization strategy introduced with the Krw-/AbfG will not enforce an environmentally sound waste business, additional instruments must be used to, for example, increasingly put product responsibility into concrete terms in order to achieve market economy powers for environmental solutions. With the withdrawal of the state from the actual waste management, an ecological framework is necessary which can secure the environmental control of privatized waste streams; at present this has only been partially realized. The current application of TASi is still limited and the insufficiently developed guidelines on product responsibility (except in the case of packaging waste) are leading to considerable uncertainties concerning the orientation of the waste regime and moreover the effects on the environment. The privatization of waste streams means that a change is necessary in regulations concerned with the waste sector. Until now, the handling of waste materials was more or less subject to the public regime and thus privatization of waste streams meant that a change in approach was necessary which recognizes that the composition and the hazardous potential of waste are increasingly influenced by the actual product characteristics. This will now mean that prevailing conditions must be achieved that guarantee that along with clearly increased efficiency, ecological issues must also be taken into consideration. A waste policy which is also a product policy can only be justified under national initiative on the basis of scientific findings in order to be accepted by the European Court. With the enacting of EU measures, independent national initiatives will no longer be possible within a European regulatory framework. A conflict between the national and supranational instances is already visible which is a conflict concerning regulatory power. In this respect, a decision was reached in Germany in 1998: justified by the principle that federal jurisdiction overrules lower legislative levels, the Federal Constitutional Court forbade local packa-

Synthesis table

	Regulations on MSW	Objectives	Responsibility	Measures applied	Factors of dissatisfaction	Institutional changes
1960s	Federal Epidemic Act; Water Act; Local ordinances	Public hygiene as a sovereign task	Municipalities as WDA	Compulsory usage of public services; to some extent free-of-charge	50 000 dump sites, regional spill overs, non-action of LAs	LAGA, Central Health Authority
1970s	Waste Disposal Act	Economies of scale, sanitary landfilling	Handing over WDA to districts, Federal jurisdiction	In addition compulsory usage of facilities, uniform standards	WDPs could not be enforced, 'waste tourism'	Environmental Programme, institutional shift of competencies
1980s	Act on the Avoidance and Disposal of Waste	Environmental hygiene, recovery is preferred	Districts, WDP by *Länder*, Divided legislative power	Voluntary agreements, 'golden restraints'	Still elimination as a preferred option, NIMBY, perceived scarcity of future capacities	Ministry on Environmental Issues, further centralization
1990s	Packaging Ordinance, TI on Waste from Human Settlements, Closed Substance Cycle and Waste Management Act	Differentiated 'prevention ladder', reuse and recycling, Proximity and self-sufficiency	Extended product stewardship, Deregulation of waste processing	Top-down measures, standards on landfill sites and on landfills such as thermal pre-treatment	Free trade issue Inefficiency Incompatibility of techniques Clean technologies underused Source reduction not encouraged effectively	Centrally claimed standards, Dual system of municipal solid waste management

Note: WDA, waste disposal authority; WDP, waste disposal plans; LA, local authorities ging taxes on non-returnables that had been introduced by several German municipalities. Local room to enforce source reduction effectively is restricted by the presence of the Packaging Ordinance.

In view of a single European market, increasing competencies will now be re-centralised in waste policy. Above all, the issue of determining product responsibility and a meaningful framework for the control of waste streams will be looked at from the point of European environmental policy. This already exists for those waste streams whose commercial value is rather high and are ultimately subject to the basic principle of free trade in the eyes of the European Court. It is inevitable that control of waste streams by the nation-states will either be impossible or limited. This is especially true for sorted household waste. The question as to how marketable secondaries can be separated from waste subjected to restrictions, in order to avoid possible spill-overs through supposed reuse abroad, is a central issue for the European legislative body.

Notes

1. We would like to thank Ceridwen Rowlands for helping us with her comments and corrections.
2. A requirement that the disposal of waste had to be carried out by local authorities without any danger to human health (§ 12 I Federal Epidemic Act of 1961); Requirements for disposal facilities (Articles 5 II No. 4 & 9 I No. 14 Federal Building Act and § 16 Industrial code [*Gewerbeordnung*]); Protection of surface waters and groundwater from pollution by waste disposal sites (Articles 26 II & 34 Water Act of 1957).
3. Free of charge at the beginning of the 20th century.
4. Materialien zum Umweltprogramm 1971, Schriftenreihe des BMI, vol. 1.
5. The IPA, founded in 1952, was a co-operation of individual members of the *Länder* and federal parliaments.
6. The task was delegated to the Institute of Water, Soil and Air Hygiene in order to use existing structures to solve the ongoing problems.
7. BT-Drs. VI/2710 and Materialien zum Umweltprogramm 1971. Schriftenreihe des BMI, vol. 1.
8. In Germany, there are two legislators. Their co-operation is similar to the division of power between Council and Parliament in the legislative procedure. The German Federal Council as one of the two legislators represents the *Länders'* interests.
9. The Mediation Committee is similar to the European legislative practice resolving conflicts between the two houses.
10. Article 74, No. 24 Constitutional Law.
11. See note 2.
12. The results are mainly based on Müller (1978) and Eckert (1985).
13. Kaimer and Schade (1996), p. 32.
14. The dominant criteria were the direct, short-term costs, whereas social costs and benefits from composting had been ignored.
15. Cord-Landwehr (1994), p. 18.
16. Thomé-Kozmiensky (1997), Vorlesungsskript Abfallwirtschaft, Technical University Berlin, p. 14.
17. Benzler et al., p. 223.
18. Spies (1994), p. 270.
19. The coalition of Liberals and Conservatives justified this absence of a norm by the already existing norms in § 5 I No. 3 Federal Pollution Control Act, which had been amended in 1986.

20. The obligation to recover waste is to be met to the extent that this is technically feasible and economically reasonable, especially when a market exists, or can be created, for an extracted substance or for extracted energy. See e.g. Köller (1996), p. 530 f.
21. Spies (1994), p. 270.
22. Last, but not least, the Bavarian Land Government filed a Federal Council legislative initiative amending the Waste Management Act in 1990.
23. Spies (1994), p. 268.
24. Schmitt-Tegge (1997).
25. The subsequent increase in the use of returnable bottles in 1982 was interpreted as a great success until in 1984 this proportion dropped again, leading to a search for the true cause of the 1982 increase: it only happened because of a very good apple harvest, followed by increased production of apple juice, which was mainly filled in refillables.
26. Meanwhile, the Packaging Ordinance replaced the disputed Deposit Ordinance and the conflict between Germany and the European Commission had been settled as there are no longer mandatory deposits in effect.
27. See also Verstey and Wendenburg (1994), p. 836 f.
28. A nationwide citizens' movement (the better waste-concept) addressed the Environmental Ministry insisting on public demands to encourage the avoidance and recycling of plastic effectively. Environmental groups went into action and unpacked products in stores and supermarkets, actions that gained sympathy and approval from the public and news media.
29. Spies (1994), p. 280.
30. 90% for beer and mineral water, 80% for fizzy drinks, 50% for wine, 35% for juice.
31. Hansmeyer (1987), p. 262.
32. See Spies (1994), p. 273.
33. Kursawa-Stucke *et al.* (1994), p. 46.
34. Lambsdorff (1990).
35 An exception is made for milk (§ 9) where the multiple use should have a percentage of at least 17%.
36. The relative proportion of packaging's value related to product's value is the highest in the case of food supplies. See Schlömer (1992), p. 254.
37. Schlömer (1992), p. 253.
38. Spies (1994), p. 287.
39. ibid., p. 288.
40. Whereas it was only thought to internalize the costs caused by different packaging materials.
41. The planned individual guarantees (each single enterprise) for the packaging material failed due to the resistance of industry.
42. In case of composites for example, the respective guarantor institutionalized a sharing of financial burdens according to market shares.
43. Under certain circumstances, the thermal value of municipal solid waste complies with the first condition and could therefore be incinerated – defined as a recovery mode.
44. Printed papers of the Federal Council no. 528/90 and 529/90.
45. Rümmele (1997). p. 3.
46. To ensure a regular utilization of generated composts market analysis with sale-concept and intended marketing structure have to be carried out.
47. *Länderarbeitsgemeinschaft Abfall.*
48. The cut down on costs for reduced landfill capacity not incorporated. Stehr (1996).
49. Experts' Committee on Environmental Questions (1991), p. 169.
50. e.g. Abfallwirtschaftsjournal 1-2/96.
51. Confidential paper prepared for Ministry of Environment.
52. SRU (1996), p. 163.
53. See Umwelt kommunale ökologische Briefe 11/1997, p. 4.
54. These costs are without proceeds of energy recovery. Umwelt 10/1996, p. 338–340.
55. These figures are reported in Umwelt 10/1996, p. 338–340.
56. Interview with Dr. Bially 21.6.97, DSD speaker.

57. Hecht and Werbeck (1995), p. 61 f.
58. Punkt 3/98, p. 6.
59. For the figures on weight see Dicke and Neu (1996).
60. Ewringmann *et al.* (1995), p. 12.
61. Interview with Dr. Eisenblätter (Federal Environmental Agency).
62. The actual trend of decreasing prices is mainly due to the fact that waste from industry shrank dramatically to 70%. Reported in Umwelt 5/1997, p. 30.
63. See for example UmweltMagazin 3/1997, p. 38, and UTA 1/1997, p. 25–27.
64. Example *Rhein-Sieg-Kreis* in Handbuch Müll und Abfall Kap. 1511, p. 17 Lfg. 9./96.
65. Behrens and Maydell (1998).
66. Dicke and Neu (1996), Annex.
67. Römer (1994), p. 83.
68. Umweltbundesamt (1997), p. 436.
69. Feess-Dörr *et al.* (1991), p. 55.
70. Umweltbundesamt (1997).
71. Examples are to be found in Staudt *et al.* (1997), p. 55. The fact is also illustrated by a decrease in specific packaging weight of non-returnables although beverage consumption increased. See Ewringmann *et al.* (1995), p. 20.
72. According to the Association for Research on Packaging Markets (*Gesellschaft für Verpakungsmarktforschung mbH* GVM).
73. The decrease of total packaging use in the year 1996 was due to the continuous decrease of tertiary and secondary packaging.
74. Due to the growing complexity and cost-intensity, direct public operation (*Regiebetriebe*) was replaced by outsourcing of services (*Eigenbetriebe*) in the 1980s.
75. Bundestagsdrucksache 12/6685.
76. Bundestagsdrucksache 12/6685, p. 13.
77. RWE is directly or indirectly connected with about 200 waste management companies; VEW is directly or indirectly connected with about 110 companies. See Handelsblatt vom 20/9/1993.
78. The predecessor VGK was not able to supply sufficient capacities. As a reaction the DKR was founded in November 1993.
79. The bank consortium which had taken over the industrial side of the DKR has good relations with the large disposal enterprises such as RWE.
80. Press release 26th January 1996.
81. UBA (1997), p. 446.
82. UBA (1997), p. 438.
83. Some of the WDAs now co-operate with municipalities abroad in order to overcome non-action. Examples are the co-operation between Coesfeld/Hengelo and Ortenaukreis/Strasbourg.
84. See Umwelt kommunale ökologische Briefe Nr. 11/1997, p. 4.
85. See Schmidt (1997), p. 442 f.
86. Interview with K.-J. Neuhaus (Präsident des VKS) documented in Entsorga 4/1997, p. 16.
87. Examples are the export of plastics to the Czech Republic in January 1992 due to lack of adequate recycling facilities and the scandal about plastics which appeared on French landfill sites in July/August 1992. See Schlömer (1992), p. 263.
88. The virgin material has been one ECU per kg cheaper than secondary material.
89. Sorted household waste belongs to the green list.
90. Besides bakeries, butchers, there are about 40 big resistant enterprises riding free. From the point of view of an individual producer, it is rational to trust that competitors will make contracts which will to a certain extent ensure consistency with the required quotas. The rational producer will benefit from the exemption from individual obligations to take back without contributing to the system. See Hecht and Werbeck (1995), p. 69.
91. According to Gunnar Sohn, DSD speaker.
92. Staudt *et al.* (1997), p. 152.

93. After having paid off the credits the disposal enterprises (especially, RWE, VEW and Rethmann groups) should leave the board of directors required by the Federal Cartel Office.
94. Michaelis (1993), p. 75.
95. The higher the contribution of each member of the group to reach the common goal, the stronger the cohesion of the interest group according to the theory of collective behaviour. For a game theoretic analysis of this problem of stability see Klepper and Michaelis (1992).
96. This is already a common practice in Bavaria, where incineration facilities face residual waste without heating value.
97. For the figures on chemical industry, see Schlömer (1992), S. 259–261.
98. German steel industry blamed the Bremer Klöckner Werke for having solved the main problem of the plastic industry which is the disposal of their waste. In their viewpoint the "reconciliation between steel and plastic industry" would bring about a considerable advantage for the plastic industry in the competition between packaging materials.
99. The facility EING is sited at Gescher. See Punkt 5/97, p. 4.
100. Kursawa-Stucke *et al.* (1994), S 31 f.
101. An example is the information of DSD to found an enterprise called EVA (intention expressed in a personal letter); however, EVA was founded in 1997 by Interseroh at Vienna.

Bibliography

Behrens, F. and O. Maydell, 1998, Analyse der Kostenstruktur der kommunalen Abfallentsorgung, UBA-Forschungsbericht, 98-040.

Benzler, G., M. Halsstrick-Schwenk, P. Klemmer and K. Löbbe, 1995, Wettbewerbskonformität von Rücknahmeverpflichtungen im Abfallbereich. Untersuchungen des Rheinisch-Westfälischen Instituts für Wirtschaftsforschung, Heft 17.

Brüggemann, A., 1996, Einstieg in die Kreislaufwirtschaft - Ausstieg aus der Wettbewerbswirtschaft? Wettbewerbliche Probleme von Rücknahmeverpflichtungen im Abfallbereich dargestellt am Beispiel der Verpackungsverordnung. Diplomarbeit an der Universität zu Oldenburg, Fachbereich Wirtschaftswissenschaften.

Bühner, A., 1997, Gewerbeabfälle privat entsorgen. In Umwelt kommunale ökologische Briefe Nr. 9, p. 17–18.

Bundesinnenministerium, 1971, Materialien zum Umweltprogramm 1971. Schriftenreihe des BMI, vol. 1.

Bundesministerium für Umwelt, Naturschutz und Reaktorsicherheit, 1990, Umweltbericht Mai 1990.

–, 1993, Technische Anleitung Siedlungsabfall. Informationen des Umweltministeriums Bonn.

–, 1995, Pressemitteilung 51/95 v. 13.11.1995.

Bundesratsdrucksache, 1990, Printed papers no. 528/90 and 529/90.

Bundestagsdrucksache VI/2710: Umweltprogramm 1971 der Bundesregierung.

Bundestagsdrucksache VI/2401: Abfallbeseitigungsgesetz.

Bundestagsdrucksache 12/6685: Entwicklung der bundesdeutschen Entsorgungswirtschaft. Sozialversicherungspflichtig Beschäftigte in der Straßenreinigung, Müllabfuhr, Abwasserbeseitigung und ähnlichen hygienischen Einrichtungen, früheres Bundesgebiet.

Bünemannn, A. and G. Rachut, 1993, Der Grüne Punkt – Eine Versuchung der Wirtschaft.

Clemens, C., 1994, Entsorgung von Siedlungsabfällen – Konfliktfelder zwischen Politik und Wirtschaft.

Cord-Landwehr, K., 1994, Einführung in die Abfallwirtschaft.

Dicke, H. and A. Neu, 1996, Nationale Umweltpolitik im Konflikt mit dem Europäischen Binnenmarkt? Das Beispiel der deutschen Verpackungsgesetzgebung. Kieler Arbeitspapier Nr. 761.

Doedens, H., 1982, Die Zuständigkeit der Landkreise für die Abfallbeseitigung.

Duales System Deutschland: Geschäftsberichte 1993–1997.

Eckert, R., 1985, Die Entwicklung des Abfallrechts. In: NVwZ, Heft 6, p. 388–398.

Emberger, J., 1997, Die Abfallwirtschaft braucht Planungssicherheit. In: Umwelt, Heft 5.

Emslander, T. and K. Morasch, 1996, Verpackungsverordnung und Duales Entsorgungssystem – Eine spieltheoretische Analyse. In: ZfU, Heft 2, S. 209–225.

Ewringmann, D., B. Linscheidt, A. Muno and J. Schuckmann, 1995, Ökonomische und umweltpolitische Beurteilung einer Pfandpflicht bei Einweggetränkeverpackungen. UBA Forschungsbericht 109 04 005.

Feess-Dörr, E., U. Steger and P. Weihrauch, 1991, Muß Verpackung Abfall sein? Strategie zur Reduktion der Umweltbelastungen durch Einwegverpackungen.

Gaßner, H.W., 1997, Ablagerung biologisch mechanisch vorbehandelter Abfälle nach dem 1. Juni 1995. Handlungsspielräume der TA Siedlungsabfall. In: Müll und Abfall, Heft 5.

Haltiner, E., 1997, Konzeptlosigkeit sondergleichen. Die Frage Wieviele Anlagen braucht das Land? ist eine Provokation. In: UmweltMagazin, Heft 5.

Handbuch Müll und Abfall: Kap. 1407, p. 17, Lieferung 3/97 und Kap. 1511, p. 17, Lieferung 9/96.

Hansmeyer, K.-H., 1987, Abgaben und steuerliche Instrumente in der Umweltpolitik. Wirkungsweise , Erfahrungen , Möglichkeiten. In: ZfU, Heft 3, p. 252–266.

Hecht, D. and N. Werbeck, 1995, Rücknahmeverpflichtungen als Instrument der Abfallwirtschaft – eine ökonomische Analyse am Beispiel des Dualen Systems Deutschland. In: ZfU, Heft 1, p. 49–79.

Kaimer, M. and D. Schade, Hrsg., 1996, Pilotstudie Bewertung der thermischen Abfallbehandlung. Planung, Bau und Betrieb von thermischen Restabfallbehandlungsanlagen; Arbeitsbericht.

Klepper, G. and P. Michaelis, 1992, Will the "Dual System" manage packaging waste? Institut für Weltwirtschaft, Kieler Arbeitspapier Nr. 503.

Köller, H., 1996, KrW-/AbfG. Reihe Abfallwirtschaft in Forschung und Praxis, Bd. 77.

Kursawa-Stucke, H.-J., N. Liebert and A. Jensen, 1994, Der Grüne Punkt und der Recyclingschwindel – Abfallwirtschaft in der Krise.

Lambsdorff, O. Graf, 1990, Duale Abfallwirtschaft statt Markteingriffe. Plädoyer für eine privatwirtschaftliche Lösung des Hausmüllproblems. in Handelsblatt v. 5.1.1990.

Lauer-Kirschbaum, T., 1996, Abfall und die Strukturen seiner Bewältigung. in Blanke, B., W. Lamping, T. Lauer-Kirschbaum and M. Plaß, Der Strukturwandel der Abfallpolitik in Niedersachsen.

Lehmann, M., 1996, Verpackungswahl und Netzwerkexternalitäten. Zur Effizienz von Verpakungsabgaben und DSD-Gebühren. In: ZfU, Heft 2, S. 227–241.

Michaelis, P., 1993, Ökonomische Aspekte der Abfallgesetzgebung.

Müller, A., 1978, Handlungsrationalitäten bei der Politikimplementation auf kommunaler Ebene. In: Mayntz, R., Hg., Implementationsprobleme kommunaler Umweltpolitik.

N.N., 1996, Mehrweganteil bei Getränken 1995 weiter auf hohem Niveau. In: Umwelt Nr. 12/96 p. 437–438.

N.N., 1997, Wieviele MVA's braucht das Land? In Umwelt kommunale ökologische Briefe Nr. 11 p. 4.

N.N., 1996, Abfallwirtschaft im Wandel. In: Umwelttechnik/Sonderausgabe, p. 10–23.

N.N., 1996, Gesamtkosten für die Behandlung von Siedlungsabfällen. In: Umwelt, p. 338–340.

N.N., 1997, Mehr Freiheit ohne Schonraum. Interview mit K-J.Neuhaus Präsident des VKS. In: Entsorga Heft 4, p. 16.

Punkt: a publication of the DSD.

Reh, H., 1997, Überlassungspflichten für Abfälle. In: Umwelt kommunale ökologische Briefe, Nr. 5, p.17–18.

Römer, A., 1994, Weg von der Wegwerfgesellschaft? Einige Anmerkungen zur Verpackungsverordnung und zum Dualen System. In: Zeitschrift für angewandte Umweltforschung, Heft 1, p. 75–97.

Rümmele, S., 1997, Bedarf von Null plus X. In: UmweltMagazin, Heft 5, p. 3.

Schenkel, W., 1993, Das Duale System als Großversuch. In: ZAU, Heft 4, p. 441–444.

Schlömer, R.H., 1992, Interaktionen von Umweltbewußtsein, Umweltpolitik und emittierenden Industrien.

Schmidt, C., 1997, Auslastung von Müllverbrennungsanlagen. Probleme, Ursachen, Handlungsstrategien. In: Der Städtetag, Heft 6, p. 442.

Schmitt-Tegge, J., 1997, Die Selbstverpflichtungen im Bereich der Getränkeverpackungen von 1977 und 1982 zur Verringerung der Abfallmengen aus Getränkeverpackungen und zu verstärkten Recyclingleistungen u.a. bei Behälterglas und Weissblech. Vortrag zur "Umweltpolitischen Bedeutung von umweltbezogenen Selbstverpflichtungen", 9.1. 1977, EAP Berlin.

Spies, R., 1994, Von der Kooperation in der Abfallpolitik zur staatlich flankierten Selbstorganisation: Das Duale System. In: Staatswissenschaft und Staatspraxis, Heft 2, S. 267–302.

SRU, 1996, Die Umsetzung einer dauerhaft-umweltgerechten Entwicklung.

SRU, 1991, Abfallwirtschaft. Sondergutachten 1990.

Statistisches Bundesamt, 1996, Abfallbilanz 1993 für Deutschland; vorläufige Pressemitteilung v. 25.1.1996.

Statistisches Bundesamt, 1993 und 1998, Fachserie Umweltschutz, Reihe 1.1., Serie 19.

Staudt, E., H. Kunhenn, M. Schroll and J. Interthal, 1997, Die Verpackungsverordnung – Auswirkungen eines umweltpolitischen Großexperimentes. Innovation: Forschung und Management, Bd. 11.

Stehr, H.G., 1996, Erfahrungen mit der flächendeckenden Bioabfallkompostierung in Hessen. In: Rheinisches Institut für Ökologie Bioabfallmanagement 1996 Tagungsband.

Thomé-Kozmiensky, K.J., 1977, Vorlesungsskript Abfallwirtschaft, Technical University Berlin.

Thywissen, C., 1995, Die Abfallwirtschaft in der Bundesrepublik Deutschland. Europäische Hochschulschriften, Reihe 5, Volks- und Betriebswirtschaft, Bd. 1765.

Umweltbundesamt, 1997, Daten zur Umwelt.

Versteyl, L.A. and H. Wendenburg, 1994, Änderungen des Abfallrechts. Anmerkungen zum KrW-AbfG sowie den Gesetzen zu dem Baseler Übereinkommen. In: Neue Zeitschrift für Verwaltungsrecht Heft 9, p. 833–843.

Wey, K.G., 1982, Geschichte der Umweltpolitik.

Wuppermann, B., 1991, Verpackungsverordnung: Machtkonzentration auf Kosten der Ökologie. In: ZAU, Heft 4, p. 352–357.

CHAPTER 2

Waste management in the Netherlands

Patrick Kalders and Wim Hafkamp

The waste management regime from 1870–1970

Legislation, technologies of waste treatment and actors

Since the end of the 19th century waste has fundamentally been a public health and safety issue (Eberg, 1997, p.3). People had to be protected from waste for reasons of health and safety and the first legislation concerning waste has to be seen as constructed within this context. The legislation concerning waste has a long history. However, a clear answer to when exactly this legislation started in the Netherlands is difficult to give. Some authors state the Nuisance Act, which was adopted on 7 January 1875, was the first Dutch environmental act. Others follow the trace back to the beginning of the 19th century when the imperial directive concerning odour nuisance was adopted (Beenakkers, 1991, p.12). Although these beginnings of waste legislation, whichever act is the first, may indicate a long history of environmental concern, it does not guarantee that the concern for environmental issue in general, and waste issues in particular, has grown ever since. It is important to note that in the period 1875–1952 one of the first environmental laws, the Nuisance Act, is said to have been particularly inactive (Aalders 1984).

As far as the organization of the waste sector is concerned, in the beginning of the second half of the 19th century waste was normally collected by private companies. However, in the period 1870–1880 many municipalities did not renew their contracts with private companies (Groen, 1966, p.3). From then on most of the municipalities had their own organizations collect the waste. Besides the municipalities, national and provincial governments are also central actors in the period 1870–1970. The municipalities were responsible for the collection of waste and also managed the disposal of household waste. Rotterdam and Amsterdam built incineration facilities, but The Hague tried to transport their waste to the East of the Netherlands where it could be composted. The Hague started a negotiation with the national government regarding this transport, which led to the foundation of The Waste Removal

N. Buclet and O. Godard (eds.), Municipal Waste Management in Europe, 63–86.

Society in 1929. This co-operation resulted in two big composting facilities in Wijster and Mierlo, which were founded respectively in 1932 and 1955 (Groen, 1966, p.4). From a political point of view the co-operation between governments and between governments and intermediary organizations is typically Dutch. As a society, the Netherlands is characterized by a consensus-driven way of policy making. At that time the society was divided into several religious groups (mainly Protestants and Roman Catholic), which had to deliberate frequently in order to make policy work.

At the end of the 19th century it was normal for household waste and cess pool substances to be collected and composted. In this way a product in high demand was produced. At the time fertilizers entered the market and the demand for compost decreased drastically, most municipalities decided to switch to landfilling. Several of the bigger cities built an incineration facility: Rotterdam in 1912 and Amsterdam in 1918 (Groen, 1966, p.4). Other municipalities lacking large budgets had to keep landfilling their waste. Consequently, in 1960 the Netherlands had several incineration facilities and about 1100 landfill sites. These developments in technology of waste treatment during the period 1870–1970 confronted the Nuisance Act with its limitations. The Nuisance Act regulated disposal sites (landfilling and incineration), but as a consequence of the use of new technologies the situation changed. Because of the change in waste technology in the second half of the 20th century also other laws became related to the waste problem, particularly the Surface Water Pollution Act and the Air Pollution Act. Another relevant change at the end of the 1960s was that the constellation of actors changed: environmental NGOs entered the waste sector.

Factors of dissatisfaction

The limitations in the Nuisance Act to deal with new technologies of waste treatment and the criticism of environmental NGOs in themselves did not lead to a radical change. Several changing conditions in the second half of the 20th century provided the setting for a more radical change in the waste sector. This change in setting consisted of an increase in the scale of waste sources and corresponding effects on the environment. Several problems contributed to the general impression that a modification of the waste sector had to be made. Problems arose with both a quantity and quality nature. The problems arose as a result of the increasing level and changing pattern of production and consumption (Eberg, 1997, p.3). The quantity aspect of the problems caused by waste consisted of an increase in the volume of waste that resulted from population and economic growth, and a higher standard of living. The quality aspect arose as a result of new substances that need special treatment and careful disposal.

Particularly in the late 1960s and early 1970s, rising economic activity was conceived as having negative side-effects on the life of the Dutch population. In some places industrial activity led to pollution which was considered unacceptable by the local population, who required the government to impose stricter measures (Boender, 1985). Also in the 1960s, national and international environmental organizations as well as a broad environmental concern in society arose. Partly, these developments are the result of early signals of a broad criticism, of which 'Silent Spring' (Rachel Carsons, 1962) is perhaps one of the best known. In the period 1968–1972 many environmental publications appeared: these can be seen as the forerunners of the environmental actions that took place in the beginning of the 1970s.

The national and international environmental concern that arose during the 1960s resulted in the early 1970s in an environmentalists campaign against one-way packaging. The objective of this campaign was to make packaging an example of the much broader criticism on the throwaway society. The criticism was picked up by industry and politics and descriptions of the problems that were perceived entered the public and private sector. They marked the beginning of the collapse of the old waste regime.

The rising of a new approach – the period after 1970

The regulatory changes

Principles in the Waste Substances Act

In 1972 the Ministry of the Environment published a Memorandum of Urgency that described the environmental problems. Also, the Ministry presented an action plan. Following the memorandum several laws were developed, like the Act on Waste Substances (1977) that particularly regulated the treatment of household waste. The Memorandum of Urgency, however, was not the only driving force behind the development of this new legislation. It also was pushed by a number of incidents in other countries, which focused attention on waste as an important issue. Special attention was given to industrial waste. Incidents on the international level intensified the concern and gained much attention, such as the Seveso case (1976), and the consequential disappearance of Seveso clean-up waste into other countries.

In the Waste Substances Act (1977) waste was defined as: "all substances, preparations or other products of which the beholder – in perspective of disposal – wants or needs to dispose of". This definition leaves open the opportunity for beneficial use, enabling trade in wastes as a product.

In the Waste Substances Act, three principles were laid down for implementation in practice:

1. The proximity principle; wastes should be treated close to the site where they are generated.

2. The self-sufficiency principle; provinces should be able on their own behalf to treat their wastes in a proper way.

3. The efficiency goal; waste should be treated in the most efficient and appropriate way.

The proximity principle and the self-sufficiency principle draw on the objective to prevent risky and non-controllable national and international transport of wastes. The events after the Seveso incident in Italy, whereby wastes disappeared untraceably before showing up in various EU countries, had their influence on the shaping of this principle.

One of the central elements in Dutch waste management policy was the adoption by parliament of the prevention priority ladder in 1978 (Figure 1). As a consequence a hierarchy in waste management was established, putting pollution prevention at the top and landslide disposal at the bottom. The background behind this hierarchy was two-fold. Firstly, the limitations in the Netherlands for continuing landfilling, and secondly, the assumption that it was more efficient to put increased priority in waste prevention.

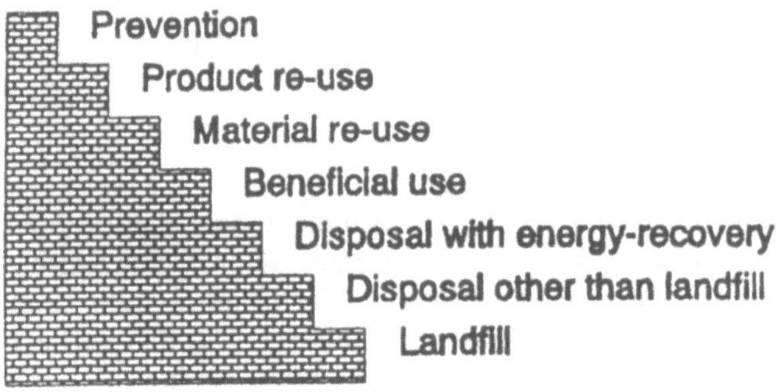

Figure 1. Prevention priority ladder

Focus on waste planning: policy reforms through incidents

During the 1980s several incidents further shaped policy by focusing the interest on the societal importance of a well established organization for waste management. In 1982 illegally dumped chemical wastes were found under a newly built residential area, Lekkerkerk, after residents complained of health problems. The wastes were excavated and removed. This was a costly operation, and the prices of the real estate declined sharply in that area afterwards. Several other sites were found around that time (e.g. Dordrecht, Gouderak).

One of the consequences of these incidents was more rigour in waste planning and control of handling and transport. In the mid-1980s, on a national and provincial level, governments started making more extensive planning for waste management (waste distribution plans). Also, in the mid-1980s the government initiated some studies on the dispersion of dioxin as a consequence of incineration and landfilling. As a consequence a guideline for incineration of wastes was issued in 1985. The dioxin issue did not regain great attention until the end of the 1980s, when it was shown that dairy products and meat products in the industrialized surroundings were contaminated with dioxins way above norms. This discovery and its consequences led to new guidelines as well as the temporary closure of several incinerators. As a result, incineration installations were put under a stricter control and were obliged to install the latest technology before they were allowed to operate. Another consequence from this event was that sensitivity and attention was gained in the public area towards incineration as an option for waste treatment.

Objectives and chosen instruments: the focus on packaging waste and covenants
The 1980s are also characterized by other developments. One important development was the development of the 'target group' approach in the national environmental policy. The Dutch government's target group policy was introduced in 1986, in order to more efficiently pursue environmental policy goals for specific groups. This policy was developed as a response to two trends. First, a trend towards deregulation in government; second, and more importantly, in several evaluations it turned out that the results of command-and-control regulation (e.g. in the form of permitting) were disappointing. By deliberating directly with the affected interest groups in society, support for a more powerful implementation was expected to increase. In addition, through a combination of 'carrot' and 'stick', voluntary and enforceable elements, policy goals were expected to be attained more efficiently as well. In total, 12 target groups were identified. In Table 1 some of the themes and target groups are presented. For the sectors within these target groups, goals were negotiated with the respective sectors. Such sector agreements took the form of intention declarations, also referred to as covenants.

Table 1. Themes and target groups

Themes	Target Groups
Climate change	Agriculture
Acidification	Traffic and transport
Eutrophication	Industry and refineries
Dispersion of toxic substances	Gas and electricity supply
Disposal of waste	Building trade
Disturbance	Consumer and retail trade
Dehydration	Environmental trade
Squandering of resources	Research and education
	Societal organization

Source: The Netherlands, OECD Environmental Performance Reviews (1995, p.34)

Next to the development of the target group policy, in the mid-1980s the Ministry of Environment started a process of investigation towards the existing waste streams. This led to the identification of 30 priority waste streams (streams and volumes for 1990 are listed in Table 2), based upon considerable negotiation and debate with the parties involved, industry, public authorities. The list of the first 29 priority waste streams was published in 1988 in the Memorandum on priority waste streams. For these waste streams, reduction goals were set in the National Environmental Policy Plan and the National Environmental Policy Plan Plus, that appeared respectively in 1989 and 1991. In negotiation with the target groups and sectors involved a process of covenant negotiation for all these specific streams was set up. The starting point was that, if satisfying covenants with one or more sectors for a certain stream could not be concluded, regulation would be designed and implemented for that sector. Covenants have not been reached in all sectors. For the sector involved in the production and selling of batteries, the negotiations for a covenant did not lead to an agreement. Instead a regulation was set up and implemented.

In the packaging covenant, two elements that are related to the principle of 'producer responsibility' were introduced. First, the element of 'chain responsibility', which means that the responsibility for the environmental effects of production is shared among the actors in the product chain. In the case of the packaging covenant, this means that the packaging manufacturers, the food and electronics producers, the recyclers and the government are considered as the responsible actors for dealing with the wastes arising from the use of packaging. This principle also represents an important structural change in policy. Instead of focusing on the environmental effects that one isolated producer or sector makes, attention is given to the columns of companies, which are tied to each other with respect to manufacturing of a product or a range of products. In this way, the effects of costly adaptations can be shared among the companies within a column. Also, in this way, more efficient solutions are encouraged, to prevent isolated changes with only one of the partners that might lead to considerable negative effects on other parts of the chain. Through a chain management approach the sharing and co-ordination of effects is encouraged, and also through the joint thinking process, a wider range of technical solutions is generated to choose from.

The second element is the life cycle principle, and the application of life cycle analysis (LCA), meaning that all environmental effects of production in all phases of the life cycle of a product are considered. This is the principle, which connects the packaging covenant as an instrument of waste minimization to product policy. Through pressure on the reduction of packaging waste by reconsidering design through stimulating life cycle assessments (in addition to take-back schemes), this link is firmly established. By this mechanism, it was tried to prevent situations, in which, in order to reduce the wastes at the end, solutions would be chosen that would cause other environmental problems or that would cause environmental problems in other phases of the life cycle of the product (either upstream or downstream).

Table 2. Volumes of priority waste streams in thousands of tonnes

Dredging materials (dry basis)	24 000
Manure (surplus)	17 500
Construction and demolition waste	12 400
Household waste	5 494
Sewage sludge	3 566
Packaging waste	3 200*
Commercial waste	2 873
Phosphoric acid gypsum	2 300
Shipping waste	2 278
Waste paper/cardboard	2 200
Street waste, market waste	1 300
Power plant ash	1 273
Municipal waste incineration residue	750
Bulk household waste	696
Polluted soil	630
Plastic waste	540
Scrap Vehicles	494
Glass (non-refillable)	430
Jarosite	220
Shredder waste	140
Hospital waste	128
Blasting grid	125
Scrap in household waste	125
Oxylime sludge	124
Waste oil	101
Car tyres	81
Halogenated hydrocarbons	55
Paint waste/spray painting waste	20
Pickling acid	8
Batteries	4

Note: some substances are counted in more than one stream

*1991

Source: VROM, CBS

The chain responsibility and life cycle principles are closely connected. Each of the producers involved in the chain is involved at another stage of the life cycle of the product. Through the life cycle perspective, the focus in the covenant is particularly also on upstream processes (e.g. redesign, substitution) in addition to downstream processes (e.g. product reuse).

The life cycle and chain responsibility principles were not only applied in the packaging covenant. Also in the national waste strategy, issued in 1995, in which goals were set for the prevention, reuse and recycling and treatment of wastes until 2010, they are important starting principles (AAO, 1995; see Table 3). Included in the waste strategy are specific programmes for the prevention of the generation of wastes from a life cycle perspective.

Table 3. Targets of national waste policy

	1990	2000	2010
Prevention	0	6	5
Recycling	30	40	40
Incineration	3	9	11
Landfill/discharge	14	5	4
Total	47*	60	60

*Excluding phosphoric acid gypsum and waste disposed of on-site, which are included in the total waste generation figure of 52 million tonnes.

Source: VROM

In the second packaging covenant, which will be concluded as a consequence of the European packaging waste directive, higher goals are being set for reuse and recycling in order to limit the household waste stream. This is partly in response to the increase in household waste.

Organization of the regime

The structure of waste management

In the 980s and 1990s a new structure of waste management was set up: management of household wastes was delegated by the national government to several regional bodies and the municipalities. A variety of public and public/private organizations on different levels is now responsible for different aspects of waste policy (Figure 2).

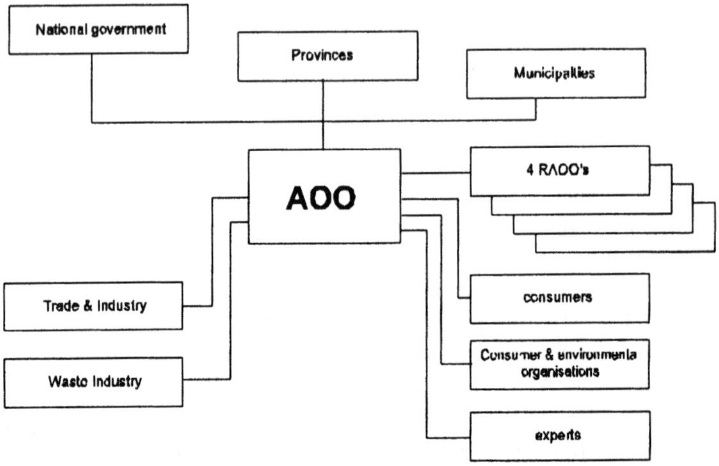

Figure 2. Structure of waste management

The national government develops and sets the national policy for the management of wastes and takes care of the transposition and implementation of EU directives. The AOO, the waste management deliberation body, established in 1990, is the main co-ordinating organization for waste management. The AOO is in charge of the development and implementation of waste strategy, and contributes to waste planning. It acts as a deliberation platform for all main interests groups with respect to waste planning. These include the Ministry of the Environment, the Netherlands Association of Municipalities, the Netherlands Association of Provinces, as well as a representation of the Environmental and Consumer organizations. Additionally, the AOO co-ordinates the work of the four regional waste management councils (i.e. RAOOs). Each of these regional waste management councils takes care of planning and co-ordination of waste streams in their regions.

Monitoring of the generation of household wastes is generally done yearly by RIVM, the national institute of Public Health and Environment. Through estimating the composition of the waste stream by at random tests in landfills, the volumes of the wastes are determined by monitoring a number of representative regions and extrapolating their volumes to the national level. In addition, for some waste streams, in household wastes, input measurements are also carried out. For example, for packaging, the packaging for households that is to be introduced newly into the market is estimated yearly by an independent consultancy.

The provinces have responsibility for the disposal of non-hazardous wastes and develop plans that follow the goal attainment for various waste streams on the national level. The municipalities are responsible for the collection and disposal of household waste.

The waste management deliberation platforms (AOO and RAOOs) operate autonomously, occasionally taking the actions in the 'discretionary' space that the national government leaves them. They operate the key function of co-ordination, and have a considerable interest in maintaining their scope of interest; as they act on their level almost as policy making bodies. Due to the fact however, that they are rather 'organically grown' they occasionally have overlapping influence fields and competing authorities. This formal overlap is mostly overcome by joint planning.

A particular feature of the institutional structure of waste management in the Netherlands is that the public sector is required to prepare treatment and treatment facilities for municipal waste as well as for waste from industry, excluding hazardous waste. At the same time, recycling operations are private, while also many joint public/private ventures exist with respect to treatment facilities and collection. The over-ruling power in treatment however, is on the side of government, who in many different forms is the major owner of most waste treatment installation. Only one of the incinerators in the Netherlands is completely privately owned.

The interwovenness of the public and private sector on the one hand may combine public and private qualities and interests, on the other hand it may

Table 4. Tasks of the different administrative bodies

National Ministry	Provincial level	Municipalities	General waste stream deliberation platform (AOO)	Waste deliberation platforms (RAOOs)
Conceptual steering on policy guidelines	Planning infrastructure permitting	Co-implementation policy provinces	Development and publication of the 10-year waste stream programme	Co-ordination waste stream capacity between provinces
Operational reformulation of municipal waste requirements to installations	Steering waste-streams Reporting, in- and export policy provincial borders	Collection of wastes	Scenario's supply of waste streams	Agreements between provinces on the allocation of wastes
Yearly plans	Policy for prevention and reuse	Involvement citizens	Planning treatment capacity incineration, disposal and treatment of organic wastes	
Hazardous wastes	Involvement management and control waste treatment facilities		Agreements on the management of waste streams between the authorities	
Im- and export policy	Involvement regional/ provincial waste management organisations			
Efficiency check Elimination hazardous wastes.	Multi-year plan hazardous wastes			

lead to non-transparent decision-making, leading to from a policy perspective unwanted results. The result may be negative side effects on both the environmental effectiveness and the financial management of waste management practice. Additional factors in relation to the interwovenness, which justify priority for increased transparency are:

1. The high financial interests that the various parties have, as well as the relatively large sums of money involved.
2. The situation that, in most cases, there is no clear 'customer' for landfilled, incinerated, otherwise treated wastes, so that there is not a high pressure on quality of the final product.

Elements of organization

The organization of the waste regime is made up of five elements. A first element is the creation of a waste market. Since the beginning of the 1990s, the national market for waste management is under pressure by first an increasing call for 'opening up' the provincial borders and second, by temporary use below capacity of certain incinerators, while there is an over-all excess of combustible wastes. The general problem identified is that supply and demand on the waste market are not matched properly and that this could be solved by a facilitation of 'market forces' in waste management. Among several actors (government on different levels, deliberation bodies, pre-treatment organizations, collectors, and management of the incineration installations) this call for increased free-market stimulation is shared. 'Market creation' here means that increasingly there will be (AOO, 1996):

1. Free access to the market for supply and demand of wastes;
2. Transparency in price-setting;
3. Free movement of goods.

These conditions, in the ideal case, must be met for all the parties in the field. As there is no real clear customer for the product of the waste sector, aside from the consumers of recyclables and reusables, a completely free market in the waste management field is not considered to be possible. Norms have to be developed, and systems set up, in order to ensure the quality of the treatment/recycling of wastes and the continuity of the treatment and management process.

At the same time, on the national level, a clear task is also identified for the different steering organizations, currently involved in waste management. A national waste council will be formed, which will supervise a proper distribution of waste streams, a second element of the waste regime, over the available capacity. The waste plans by this national operating organization are developed in conjunction with provinces and municipalities. Overall it seems that a menace for the system of increased market emphasis that keeps quality of the

Table 5. Current maximum and minimum incineration (I) and disposal (D) prices in EU countries (ECU/ton)

Countries	I Minimum	I Maximum	D Minimum	D Maximum
Belgium	27	41	19	32
Denmark			30	43
Germany	**47**	**122**	**15**	**73**
Greece			2	3
France				
Ireland				
Italy				
Luxembourg				
Netherlands**	**75**	**100**	**30**	**60**
Portugal	25	50	6	16
Spain	30	30	7	10
UK	17	35	7	11

Source: E. Dijkgraaf, March 1996, p. 21, and updated and adapted in July 1997 for Dutch prices (including the charges in several provinces/regions to make landfilling for combustible waste less attractive)

*For categories with missing data, no information was available or no considerable import capacity for landfilling or incineration exists.

**Prices are corrected for an average of provincial charges that are levied to keep incineration charges competitive with landfilling prices (e.g. in the province of Brabant, a charge is put on landfilling in order to prevent the disposal of combustible wastes).

– High cost countries are in bold print.

The Ministry of Agriculture was also involved, as a large amount of packaging is needed for agricultural products (e.g. foils for foods and products). In addition, this Ministry represents one of the most important economic interest groups in the Netherlands, that has considerable interest in changes in the food sector.

The Ministry of Economic Affairs, dealing with industry and technology policy was another actor involved, as it is regarded as one of the key actors with respect to the relations with industry. It is also one of the main actors with respect to taking into account the integration of environmental aspects in economic policy.

The various industries involved were represented by the Foundation for Packaging and the Environment (FPE), a foundation representing corporations, including packaging manufacturers, packaging fillers, and waste handlers. The majority of the large companies involved were represented in this foundation, which has about 250 members, including a large majority of the market for packaging and food products in the Netherlands. This foundation has existed since 1971 and was set up by industry in order to be better able to respond to a EU directive on the recycling of glass. Since the end of the 1980s, the foundation has worked on the preparation of the packaging covenant, and represented the main industry associations and individual companies that were part of the packaging chain. This includes the organizations that are involved with or have an interest in putting packaging on the market in the Netherlands, either by supplying raw materials for packaging materials, by manufacturing packaging, by using that packaging, by putting packaged products on the market, or by recycling packaging.

The consumer organizations and environmental NGOs were involved in the peer review process of the life cycle analysis (LCA) and the connected market economic analysis (MEA) but were not undersigning parties in the covenant.

Upon basis of a strategic discussion that involved all the stakeholders at the end of the 1980s, studies were made for the different sectors involved with packaging about the feasibility of goals for waste reduction. These studies were conducted by the companies involved, the FPE and consultancies.

Objectives of the packaging covenant

The text of the covenant contains quantitative goals for pollution prevention, product and material reuse, as well as a number of qualitative goals. These goals were set upon the basis of an exploration of the possibilities for the different waste streams, and as a result of negotiations between the different stakeholders, and the Ministry of the Environment, which were held earlier, at the end of the 1980s. These negotiations led to the setting of goals that were considered as attainable by the parties involved.

The general goal regarding quantitative pollution prevention is to reduce the quantity of packaging waste to be landfilled by the year 2000 down to nil. A second overall goal is to reduce the amount of packaging waste generated to below the level of the reference year 1986 in the year 2000. A third goal is to reduce the quantity of packaging to be newly introduced into the market in 1994 to below the 1991 level.

With regard to qualitative prevention, the packaging chain promises to substitute environmentally hazardous elements of packaging with environmentally less hazardous forms of packaging waste by the year 2000. Beyond 2000, packaging should not contain any environmentally hazardous substances, such as heavy metals. An important part of the qualitative goals is to carry out LCAs and connected to that, MEAs of the different product alternatives identified in the LCAs. With regard to materials reuse, the packaging chain undertakes to reprocess at least 60% of the packaging materials (to as high a grade as possible, and not including energy recovery/incineration).

In addition to these total figures, subscribers of the covenant agreed recycling targets for specific materials to have been reached by the end of 1995. These include percentages as (for example) for glass 80% of the used amount of glass, and for paper and cardboard 60% of the used amount of paper and cardboard.

The extent to which these goals have been attained by the packaging chain is discussed in the analysis below of the improvement of environmental quality from a life cycle perspective.

Outcomes of the new policy

Results of the packaging covenant

So far, the experience with the covenants that were concluded is relatively satisfying, even if, theoretically, its enforceability is still under debate. Evaluations of the packaging covenant (packaging is one of the priority waste streams) that was signed in 1991 (Commission on Packaging, 1995), were positive for the use of this instrument, even although not all goals have been met. One of the most important positive experiences is that, due to the interaction and negotiation preceding the covenant, and the involvement of all parties in the product chain, more creative and efficient solutions are found in achieving the original goals, than in a situation where the government negotiates with one partner at a time.

Since 1991, the Commission on Packaging and the Environment has published yearly reports reporting on the progress of the industry that undersigned the covenant in implementing the different measures. With regard to the first general goal of the packaging covenant, the reduction of new packaging to be introduced into the ecosystem up to nil in the year 2001, this was nearly attained in 1995. Since that time packaging wastes can no longer be dumped, but only incinerated with energy recovery or otherwise recycled/reused (Ministry of Housing, Spatial Planning and the Environment, 1996). Inevitably, however, some packaging wastes are included in the relatively small amount of household waste that is still landfilled. With regard to quantitative prevention: most of the actions that the producers took, belonged to this category. Looking at the type of actions that are reported by the corporations involved, the bulk of actions was reported in 1994. The majority of these were of a preventive nature. For 1994, 190 preventive actions were reported on a total of 325 actions. Examples include: the exclusion of transport packaging, the reduction of foil packaging, and the transfer from single-use to reusable packaging.

From the available monitoring data, here the focus will be particularly on the output estimations of the RIVM (National Institute of Public Health and the Environment). The output data are based on information from the industrial and household waste streams that are offered for treatment (Table 6). The proportion of packaging in the waste stream is calculated through (yearly adapted) random sampling to determine the percentage of packaging in the waste stream. This percentage is then multiplied by the total amount of measured industrial and household waste and the amount of packaging material waste arising from consumer use is added.

The term household wastes here refers to all wastes, collected from households that is dealt with through landfilling/separation, recycling or incineration (with or without energy recovery) by the municipal authorities (Table 7).

The term industrial wastes refers to all wastes, hazardous and non-hazardous that comes from an industrial sources (several categories specified

Table 6. Total amount of packaging waste generated in the Netherlands in kilotonnes (1991–1994), including material reuse, compared to 1994 goals

Material	Plastics	Paper/ cardboard	Glass (disposable)	Ferros	Aluminium	Total
Year						
1991	645	1688	558	263	46	3201
1992	647	1658	523	325	49	3202
1993	538	1500	504	201	18	2761
1994	613	1415	463	189	19	2699
Goals 1994	<645	<1688	<558	<263	<46	<3201

Source: National Institute of Public Health and the Environment, 'Analyse Verpakkingsafval 1994', report nr. 776201021, p.8

Table 7. Amount of packaging waste in kilotonnes generated by households (1991–1994), including material reuse

Materials	Plastics	Paper/ cardboard	Glass (disposable)	Ferros	Aluminium	Total
1991	308	381	517	110	18	1334
1992	334	404	483	115	18	1354
1993	268	419	481	100	11	1279
1994	289	452	441	98	12	1292

Source: National Institute of Public Health and the Environment, 'Analyse Verpakkingsafval 1994, report nr. 776201021, p.8

Table 8. Amount of packaging waste in kilotonnes generated by industry in the Netherlands (1991–1994), including material reuse

Materials	Plastics	Paper/ cardboard	Glass (disposable)	Ferros	Aluminium	Total
Year						
1991	104	423	16	78	9	631
1992	80	370	15	135	12	612
1993	69	287	6	49	3	414
1994	81	229	7	37	1	355

Source: National Institute of Public Health and the Environment, 'Analyse Verpakkingsafval 1994, report nr. 776201021, p.29.

by type and number in specific government regulations), and is handled by a private or public waste handler for product reuse, material reuse, incineration (with or without energy recovery) or disposal (Table 8).

Table 9. Amount of packaging waste in kilotons generated by the services and retail sector in the Netherlands (1991–1994), including material reuse

Materials	Plastics	Paper/ cardboard	Glass	Ferros	Aluminium	Total
Year						
1991/1992[a]	179	884	25	75	19	1182
1993	147	794	17	52	4	1014
1994	189	734	15	54	6	998

Source: National Institute of Public Health and the Environment, 'Analyse Verpakkingsafval 1994, report nr. 776201021, p.29.

[a]The data from the wastes of the service sector has only been systematically and separately been collected since 1992. In total waste statistics the 1991 numbers have been estimated upon basis of the 1992 measurement, and therefore are not presented here separately from 1992.

The term service sector and retail wastes refers to all office and retail waste (paper, cardboard, offices), discarded by private enterprises, not being industrial wastes), being reused as a product, reused as a material, recycled, incinerated (with or without energy recovery), or landfilled (Table 9).

All of the targets set for packaging waste reduction in 1994 are met, following the numbers from RIVM. Now that on the covenant level the goals have been met, a closer look at the performance of the different waste generators can be taken. Tables 5–8 show that the relatively high decreases that were attained in the industrial sector were partly reversed by the increases from the service and retail sector. Furthermore, the plastics waste stream, although staying within the covenants goals, seems to show a particular growth in the service/retail and household sectors. This can be partly explained by the fact that in many cases packaging was shifted from cardboard to plastics in a stream-lining exercise. In addition, the service sector has grown considerably since the end of the 1980s (CBS, 1994).

Also, looking at the input numbers (Table 10), which basically estimate the future packaging wastes by estimating the amount of new packaging that is being put on the market, the goals of the covenant are mostly met. The exception here, however, is the plastics waste stream which largely attains the covenant goals but misses it by a mere 6%. The relatively low growth of the other waste streams is striking. For the recycling and reuse measures, most use was made of the existing channels, i.e. the municipal waste collection of the municipalities and the existing recycling infrastructures for glass, paper and cardboard recycling. All the recycling goals were met except those for plastics. A pilot project showed that plastics recycling was problematic and expensive. Studies of new methods were set up. The increased reuse and recycling goals under the new covenant were planned to be realized by the municipalities. At the end of the 1990s the debate is how this will be financed: will the municipality be responsible for its financing, or the industry?

Table 10. Input data: index numbers for changes in total waste generation (both from industry, households and service and retail sectors) expressed in deviations from the base year 1986 by type of material for the 1991–1994 period*

Materials	Plastics	Paper/ cardboard	Glass	Ferros	Aluminium
Year					
1986	100	100	100	100	100
1991	123	122	99	113	108
1992	128	124	99	118	116
1993	129	120	97	110	104
1994	131	122	96	110	100
Goals 1994	<123	<122	<99	<113	<108

Source: Commissie Verpakking en Milieu: Jaarverslag commissie verpakkingen, October 1995, p.26.

*Imported wastes are also taken into account, as all the packaging waste is included, that undergoes some form or treatment, recycling, or is landfilled in the Netherlands.

Product reuse (e.g. reuse of milkbottles) is not included in this table, as it is not considered as a waste, but nearly immediately reused as a product.

The results presented here show that the input packaging waste reduction goals set in the covenant for 1994 have been mostly attained, with exception of the rise in newly introduced plastic packaging, which rose slightly above the 1994 interim goal.

The impact on the environment of the LCA/MEA exercise cannot be easily measured. Not all alternatives that were selected as most feasible have been implemented. However, along the line of this exercise, many enterprises perceived the costs and opportunities for win–win solution in their production. Since the start of the LCA/MEA activities and the work of the covenant for the corporations, a few hundred innovative adaptations to packaging, including complete redesign of the packaging used or simply deleting certain transport packaging, have been implemented (FPE, 1992, 1993, 1994, 1995). All of these new types of packaging were either lighter and smaller, or used less composites. Most of these types of changes are listed and described in the yearly report on packaging developments, published by the FPE. No systematic study has been done that specifically assesses the total effects of these incremental changes. In any case the use of resources for producing packaging material used for packaging products in the Netherlands has gone down, as a consequence of the reduction of newly to be introduced packaging on the market.

One illustrative result of the LCAs and MEAs can be given however. One of the more elaborate LCAs and MEAs was carried out for milk packaging, which compared three main alternatives: the glass bottle, a returnable polycarbonate bottle and a plastic sachet. Both the LCAs and the MEAs showed the sachet to be the most favourable alternative. For the MEA, the micro- and meso-economic effects of the different alternatives throughout the whole chain were estimated (benefits and costs for the packaging producers, fillers and importers/retailers). Even although the sachet was identified as the most viable alternative from a quantitative business–economic perspective, in the end it was not chosen, as a qualitative study showed that the sachets would cause considerable inconvenience of the customer. The sachet would have possibilities for leaks, and on-the shelf presentation would be less. The reusable polycarbonate bottle was chosen, as it was second best from an environmental perspective, as well as from the perspective of costs for the actors involved in the milk packaging chain.

Whether the covenant in which domestic goals were set can also lead to waste reduction from an international perspective, can be partly derived from the national statistics, which include the export of newly introduced packaging material, as well as the import of packaging that becomes waste. An overall systematic evaluation cannot be done on that basis, however, as reliable, overall statistics regarding the export of packaging waste that is reused elsewhere are not available. Some sectors, e.g. the paper recycling sector, have their own systems in place to map these flows, but they are not uniform enough to be compared with other sectors or to be connected to packaging. These numbers may show up from the evaluation of the implementation of the EU directive on packaging wastes.

Correcting evolution

Factors of dissatisfaction

Although the packaging covenant led to positive results, criticism arose regarding the organization of the waste regime. The current institutional and funding structure of the Dutch waste management arena is subject to debate, for three main reasons.

First, many institutions including the Ministry of the Environment, the Municipalities, the national and regional waste councils, and the provincial administration, have overlapping responsibilities with respect to waste management. This leads to duplication and non-transparent deliberation structures, particularly where it concerns the re-organization memorandums that many of these institutions have issued.

Second, there are financial problems. There is considerable incineration capacity and several new incinerators will become available; however, much of the waste that should be incinerated is still landfilled. This means that the capacity of some incinerators is, and will be, underused, thereby leading to great financial losses for these installations (Hafkamp, 1995). On top of that many incinerators of an older date can only function properly when their capacity is fully used. In addition, municipalities where landfills are going to close will lose a traditional source of income: landfilling fees were a welcome financial extra for municipal policies. The process of closing a landfill is also quite costly, as such sites need special attention and control. Furthermore, for maintaining the remaining landfills the costs will rise, as less wastes will be landfilled, while they need the same operation and control.

Finally, a process of scale enlargement has taken place, both in landfilling and incineration. In contrast to the proximity principle, the spatial scale of the waste markets increases as waste management companies also internationalize. The EU makes efforts to shape the policy for an EU waste market, and according to OECD council acts and the Basel convention, treatment of particular wastes is not bound to borders, taking into account certain regulations.

Policy debate and new developments

The policy debate on the organization of the waste regime, which started in the mid-1990s as a result of the described movements of dissatisfaction, is divided into an 'upstream' discussion, particularly on packaging, and a 'downstream' discussion, on post-consumer waste management.

A first policy debate concerned the packaging covenant. In 1994 the European Directive on Packaging and Packaging Waste (94/62/EC) disturbed the implementation of the Dutch covenant. According to this directive each EU Member State had to implement the directive in national legislation. A covenant would not be sufficient. Negotiations about the Dutch implementation of the Directive resulted in a ministerial guideline 'Packaging and Packaging Waste' and a new covenant (VROM, 1997; Hozee, 1997, p.15). A difference from Covenant I was that the second covenant was built up by various co-covenants (e.g. covenants for glass, paper/cardboard, metals, woods, etc.). The integral covenant was signed by several actors: Ministry of Environment (on behalf of the Ministries of Economic Affairs and Agriculture, Nature and Fishery), The Federation of Dutch Industry (VNO/NCW), Association for Packaging and Environment (SVM), and the Dutch Association for Small and Medium-sized Enterprises (MKB-NL). In July 1997 the Dutch regulation came into effect and the Covenant II was signed.

In addition, by the drafting of the European Directive on Packaging and Packaging Waste the Dutch Covenant on Packaging lost its central status. EU regulation requires implementation through national legislation. Consequently, not only a new covenant but also a new national directive was discussed. By the introduction of the Dutch Directive on Packaging and Packaging Waste the situation of industry changed significantly. The second Covenant on Packaging now had obtained a legal safety net in the sense that non-compliance with the covenant lead to legal control in accordance to the national regulation. If a company chooses not to follow the covenant it thereby places itself automatically under the jurisdiction of the new national Regulation on Packaging and Packaging Waste. In the eyes of industry the safety net will prevent strategies like free-riding behaviour. However, industry also had other reasons to comply with the Covenant II. The main reasons were the avoidance of additional costs, dispensation from legislation, a 5-year delay, and the advantages of collective lobbying and reporting, and safeguarding a free market (Hozee, 1997 p.16).

The second policy debate focuses on developments, problems and the restructuring of the waste market. In the mid-1990s the waste sector was confronted with several problems. These problems were related to certain developments in the waste sector:

1. Waste policy: the central position of the prevention ladder (as described in a memo on prevention and recycling);

2. Waste market: scale enlargement, innovations in environmental technology, increasing specialization, privatization, increasing competition, and more emphasis on efficiency (Fleurke and Huizenga, 1996, p.218).

The different interests of the parties involved, the evolving international field with respect to waste treatment and recycling, and the large-scale changes in the national structure of the market and organization urged the government in the beginning of the 1990s to take steps to reform the overall structure of Dutch waste management. As there is an increasing pressure on the waste management market, in order to create more transparency and ensure the continuity, reliability and quality of waste management according to government policy, the Ministry of the Environment installed a commission in 1996 to deal with the problematic aspects in waste management policy. This commission was called the Commission on the Future Organization of Waste Treatment (CFOWT), and consisted of representatives from the field of waste management, government and large corporations. This commission regularly consulted with various other interest organizations, including the sector organizations and the environmental interest organizations. The commission came out with a memorandum in 1996 and addressed the following policy problems (CFOWT 1996:II):

1. Waste steering;

2. Incineration and disposal of non-hazardous waste;

3. Conflicting interests;

4. Quality assurance;

5. Environmental permitting for installations and regulation regarding implementation practice;

6. Division of tasks and structure of co-operation.

On each of the policy problems mentioned above the commission proposed several recommendations. The main message of the Commission was that the proximity principle should be left, so that on a national level a better match between demand and supply of waste will be reached. A more coherent organization is necessary to be able to ensure minimum environmental standards and enforcement, while at the same time market forces are stimulated. For the latter, proper incorporation of the processes taking place at the European level is essential. This will put further demands on a better steering organization.

At the end of 1996 the Minister of Environment reacted to the advice given by the CFOWT. The Minister basically followed the advice of the commission, and put an even stronger emphasis on the liberalization of the market on a national and European level, while stressing the role of keeping minimum standards and ensuring quality and continuance. An important conclusion in European perspective the Minister made was that the aim of the waste policy would be to open the national borders to combustible waste within several years. However, for waste that is supposed to be landfilled and to be used for energy recovery the national borders will remain closed.

As a result of the recommendations of the CFOWT the Environmental Management Act will be changed. In the revised Act the regional borders, which were used to manage waste, will be opened and waste will be managed on a national level. Additionally, only one national waste control plan will be composed, the Minister of Environment will have a 'Declaration of No Objection' (concerning the permitting of incineration facilities) as a new policy instrument, and the possibility will be created to apply the instrument of tariff setting within permits for incineration facilities and landfilling sites. The new Act also anticipates the coming European market for combustible waste. However, some conditions for the international trade of waste are given. First, regarding combustible waste there is in principle no drawback on opening the national borders as soon as the surrounding countries have similar facilities (level playing field) and are willing to open their borders too. Second, concerning waste to be landfilled national self-sufficiency will be the standard, and not the European market. And thirdly, although for waste recovery a European market has already come into being, an important question that remains is how the term 'waste recovery' has to be understood.

References

Aalders, M.V.C., 1984, Industrie, milieu en wetgeving: de Hinderwet tussen symboliek en effectiviteit. Kobra, Amsterdam.

Afval Overleg Orgaan, 1995, Ten Year Waste Strategy Program 1995–2005. AOO, The Hague.

Afval Overleg Orgaan, 1996, Werking van de Afvalmarkt (functioning of the Waste Market). AOO, Utrecht.

Beenakkers, E.M.T., 1991, Aandacht van de overheid voor bodembescherming, sinds wanneer?: een literatuuronderzoek. Ministerie van Justitie, Den Haag.

Boender, C.J.B., 1985, Milieuprotest in Rijnmond: sociologische analyse van milieusolidariteit onder elites en publiek. Sijthoff Pers, Rijswijk.

CBS, 1994, Nationale Rekeningen. Den Haag. Commission on Packaging (see Commissie Verpakkingen en Milieu) (1995) [s.l.].

Commissie Verpakking en Milieu, oktober 1995, Jaarverslag Commissie Verpakkingen. [s.l.].

Dijkgraaf, E., 1996, De invloed van Regulering op de werking van de afvalmarkt en de kosten van concurrentiebeperkende regulering. Research Proposal, OCFEB, Rotterdam.

Eberg, J.W., 1997, Waste Policy and Learning: Policy Dynamics of Waste Management and Waste Incineration in The Netherlands and Bavaria. Amsterdam University Press, Amsterdam.

Fleurke, F. and F.P. Huizinga, 1996, Tussen open markt en overheidsplanning: een sturingsmodel voor afvalverbranding (Between Open Market and Policy Planning, a regulative model for waste incineration). Beleidswetenschap, 10(3), 215–238.

Groen, N.J.A., 1966, Het probleem van de afvalstoffen in het algemeen: inleiding voor de Commissie industriewater en milieuhygi·ne van de Raad van Nederlandse werkgeversbonden en de Afdeling voor gezondheidstechniek van het Koninklijk instituut van ingenieurs. Den Haag.

Hafkamp, W.A., 1995, Turbulentie en Terugval. Oratie Erasmus Universiteit Rotterdam, Rotterdam.

Hozee, M., 1997, Regulating Packaging Waste: The Dutch Story. Research Memorandum, Erasmus University Rotterdam, Rotterdam.

Ministerie van VROM, 1989, Nationaal Milieubeleidsplan. SDU, Den Haag.

Ministerie van VROM, 1991, Nationaal Milieubeleidsplan Plus, SDU, Den Haag.

National Institute of Public Health and the Environment (RIVM) Analyse Verpakkingsafval 1994. Bilthoven.

OECD, 1996, Pollution Prevention and Control, Extended Responsibility in the OECD area. Phase 1 report; OECD Environment monographs 114, OECD, Paris.

OECD, The Netherlands, 1995, OECD Environmental Performance Review, OECD, Paris.

VROM, 1997, Convenant Verpakkingen II. VROM, Den Haag.

CHAPTER 3

Municipal waste management in France

N. Buclet, C. Defeuilley, S. Lupton[1]

Introduction

The aim of this chapter is to analyse the main features of the French municipal waste regime. Our object of study will be limited to household waste, which is by far the main category of municipal waste.[2] French household waste policy has undergone important changes since 1992. Current legislation tends to promote valorisation[3] modes and to limit elimination (landfilling, incineration without energy recovery). This new regime is led by a bottom-up approach. National authorities play a role of orientation, without specifying neither the valorisation techniques to use, nor precise targets to attain. In this approach, municipalities are responsible for the translation of general orientations into technical trajectories. The sum of each local decision regarding the adoption of valorisation techniques will, therefore, depend upon the achievement of national objectives. In such a situation, the set of rules and instruments put into place at the national level influences local decision making regarding technological adoption. The institutional conditions of technical adoption by decentralized agents is therefore at the centre of the stage (Defeuilley and Godard, 1997). Can we expect a least-cost implementation of national targets, based on an efficient combination of valorisation techniques? One main condition should be satisfied: decentralized agents have to face an institutional framework providing the right economic signals (Godard, 1997). In this chapter, we will see that this is not exactly the case. A structural bias in favour of energy recovery adoption can be underlined. This is the consequence of the relative weakness of institutional schemes set to promote recycling and composting. This technical trajectory in favour of incineration calls the implementation of national goals into question. Nevertheless, at the same time, the new organization is able to develop collective learning, and adapt to changing contexts through the provision of a regular revision of targets and incentives.

This chapter is structured in three parts. The first part describes the general

N. Buclet and O. Godard (eds.), Municipal Waste Management in Europe, 87–119.

situation of the French household waste regime during the 1970s and the 1980s. The second part of the chapter deals with the regulatory and organizational changes which occurred in 1992. The third part presents the first results obtained by the new regime and analyses the recent corrections made to reduce its shortcomings. Concluding remarks follow.

The evolution of municipal waste management (1975–1990)

The need for a national legislation

Up to 1975, household waste management was characterized by a lack of control and organization, while the amount of household waste consumed increased. In 1972, the annual production of household waste per inhabitant amounted to approximately 290 kg, compared with 220 kg in 1960, an average annual increase per inhabitant of about 2% in weight. This rise was mainly due to the development of paper and plastic packaging. The quantity of household waste produced in 1972 amounted to 11 million tonnes.

In the early 1970s, household waste treatment in France was heterogeneous. Treatment plants and collection systems did not extend to the whole country: landfilling was the most common way to eliminate waste. In 1970 nearly two-thirds of the waste collected, generated by a population of 29 million inhabitants, was landfilled (GEERS, 1974). Most of these landfill sites were uncontrolled. Incineration supplied 8.5 million inhabitants (20% of the population) in 1970, 4 million of which lived in the Parisian agglomeration, where 47% of the incineration facilities were concentrated, the rest of which supplied big cities (geers, 1974). The lack of sufficient collection services and treatment units in all local communities, the increasing amount of household waste and the proliferation of uncontrolled landfilling induced public authorities to consider the establishment of an urban waste management policy with the aims of expanding collection and treatment facilities. A national regulatory framework was, therefore, set up. The law of 1975 assigned the responsibility and cost of household waste elimination to local authorities, which became obliged to provide the population with a household waste management service. The law suggested that local authorities should finance the cost of treatment and collection by setting up a tax for household waste removal. This tax system was gradually adopted by most local authorities, enabling the latter to invest in treatment facilities. This legislation marks the beginning of the concept of household waste treatment as a public service.

The situation from 1975 to 1992

With the adoption of the legislation of July 1975, local authorities started investing in treatment facilities. Four household waste treatment options were used in France: controlled landfilling, incineration, incineration with energy

Law N° 75-633, July the 15th 1975

Those who produce or are in possession of waste, in circumstances that could be noxious to human health and the environment, must provide treatment. (art. 2). concerning waste which can be harmful (art. 2), these waste can only be treated in facilities that have been approved by administration. Local authorities must ensure the treatment of household waste. Local authorities can create in this aim a special tax (art.12). This service must be ensured for the whole territory in a five-year delay. (art. 12). Only the communal service (and, should the occasion arise, authorised individual entities) can receive these waste. Waste treatment must be ensured in circumstances, which favour material and energy recovery. anred (National Agency for the Recovery and Treatment of Waste) is created. It is a public governmental institution in charge of both facilitating treatment and recovery operations, or acting in this area to satisfy public interest in the case of the lack of private or public means. anred will merge in 1992 with another national agency to create ademe (Agency for the Environment and Energy Management).

recovery and composting. The choice of treatment facilities depends on economic aspects (treatment costs) and geographical context (tonnage concerned, available sites). Elimination options (incineration and controlled landfilling) are preferred because they enable to absorb an important amount of household waste at a small cost. Landfilling is the less expensive and easiest option for local authorities. Controlled landfilling costs about 12.3 Euros/tonne in 1985, against around 24.6–26 Euros/tonne for incineration (with or without energy recovery).

Incineration expanded also after 1975, and the number of incineration units almost doubled in 3 years, between 1975 and 1978. Incineration with energy recovery developed in certain local areas (mostly big cities), where the volume of household waste to be treated was considerable and/or no landfill sites were available. Incineration with energy recovery enabled local authorities to reduce their costs in energy consumption, which represent an important percentage of their budget (4–10%), and to take advantage of economies of scale. In 1970, nine plants existed, and treated the household waste of 7.5 million inhabitants (15%). In 1985, 64 units treated the waste of 13.8 million inhabitants (25%) (ANRED, 1987).

The main objectives of the law of 1975 have been accomplished: at the end of the 1980s, 94% of the population was supplied with household waste treatment facilities, among which elimination units (incineration with energy recovery and controlled landfilling) were predominant. Uncontrolled landfilling had, therefore, been considerably reduced.

Incineration with energy recovery in Paris

Paris has a long tradition of incineration of its urban waste. The first incineration plant was established in 1912, and two other units were constructed in the following years. These plants treated approximately 1.75 million tonnes of municipal waste. In the 1950s and the 1960s, these plants were reshaped, developed and equipped with energy recovery systems. At that time, they were amongst the biggest in the world. The site at Ivry treats 800 000 tonnes of urban waste a year.

Paris and its suburbs soon chose to rest upon incineration with energy recovery for several reasons. First, the agglomeration lacked landfill sites and was very soon confronted with the problem of handling an increasing volume of waste. Second, it was obvious that the cheapest way to treat urban waste in such quantities was to build large plants, enabling to gain economies of scale. Third, Paris has developed since 1928 one of the biggest district heating in the world. This district heating is supplied by two sources: coal (for 60%) and municipal waste (for 40%) burning in the three incineration plants. The district heating need incineration plants to ensure steam production: the steam produced is sold at a high price (compared to other plants in different local contexts). Therefore, the price of incineration in Paris is far below the national average level. In 1985, the price was 10 Euros/tonne compared with 24.6–26 Euros/tonne at the national level. This difference is still true. Paris is still very bound with incineration, although some effort is made in the development of separate collection schemes (only for paper and cardboard) and sorting.

Despite its achievements, different elements brought the French waste management policy into question in the late 1980. Both endogenous and exogenous factors acted to re-orientate the French waste management system (Table 1).

Table 1. Household waste treatment facilities in 1989

Treatment mode	Tonnage	Percentage	1975–1989 trends
Landfilling	9 057 000	44.4	↗
Illegal dumping	1 600 000	7.8	↘
Incineration	2 125 000	10.4	↗
Incineration with energy recycling	5 586 000	27.4	↗
Recycling	739 000	3.6	↘
Composting	1 314 000	6.4	↘
Total	20 421 000	100	↗

Source: ADEME and GEERS

First of all, a general protest movement arose against the technical choice based on elimination. In addition to growing public concern about the noxiousness of landfilling, the population became increasingly hostile to the building of landfill sites (as the *Montchanin* example indicates). This was also true of incinerators, which are accused of emitting diverse atmospheric pollutants (heavy metals, hydrochloric acid and dioxins) which could be dangerous for human health. New demands were made upon waste management options and elimination was no longer a legitimate option. Local populations demanded facilities that may have environmental benefits: household waste became considered as a material to be valorised and not as rubbish to be dumped. This evolution induced a considerable change in public policy. Second, the French household waste system had to cope with an increasing volume of waste, which pressed on treatment capacities. Since 1960, household waste has increased by 60% (Table 2). Two elements are responsible for this rapid growth: packaging and paper (essentially newspapers and magazines). Packaging waste has increased in weight by 236% in 30 years, and now represents 33.5% of the total mass of waste produced, compared with 16.5% in 1960 (ANRED, 1991). Paper waste has risen in volume by 120% in 30 years.

Table 2. Composition of household waste in France between 1960 and 1990

	1960		1990		
	Weight (%)	Kg/inhab.	Weight (%)	Kg/inhab.	Growth rate (%)
Packaging	16.5	36	33.5	121	+236
Other papers	18	40	24.5	88	+120
Textiles	5	11	2	7	–36
Organic waste	26	57	25	90	+58
Other	34.5	76	15	54	–33
Total	100	220	100	360	+64

Source : Quirion P., 1994

This increase was confronted by a saturation, and even a shrinkage, of waste disposal services, notably due to the negative reaction of the population to, and the closing down of some sites which were not in accordance with environmental protection rules. This generated a situation in which the rhythm of dump filling increased, whereas the life expectancy of these plants decreased rapidly (ANRED, 1989). The waste disposal service, therefore, appeared to be threatened in its technical, environmental and political viability. The general protest movement against elimination sites, combined with the increase of the waste produced, led to reconsideration of the techniques in use and opened the scope of waste management schemes to new or neglected options (e.g. prevention, sorting and recycling, composting). The calling into question of landfilling as a sufficient answer to the waste problem was also influenced by pressure coming from neighbouring countries, particularly Germany. Germany was among the first countries in Europe to encourage the development of alternative techniques to elimination (i.e. recycling and re-use) with the adoption, in 1991, of the 'Töpfer' ordinance. Concerning household packaging, this ordinance obliged all industries to take back their waste and provide recycling, but allowed them to delegate this responsibility. Of course, such a delegation implies a financial contribution from the industry. High recycling objectives were set, and a contribution on different materials was imposed on the fillers, which varies exceedingly according to the materials used (see Chapter 1 in this volume).

This German legislation is perceived as a threat by the French packaging industry. In 1989, a first round of negotiations concerning a new european legislation regarding packaging waste began. The German legislation was taken as the reference for the whole European Union, which would force France to respect very demanding recycling objectives. The French industrial fillers advocated free choice in materials, a reasonable level of recycling, and did not want technological choice to be imposed. Furthermore, French packers did not want to pay as much as German industries for the packaging waste management system. Therefore the French packaging industry was urged to organize itself in order to show its own interests to the public authorities and the European Union. Hence, Germany's initiative helped to destabilize the French household waste regime, which had already been endogenously weakened. In this context, public authorities as well as the other concerned actors had to redefine the French household waste policy, starting in the early 1990s. The starting point, in 1991, came from the Ministry of the Environment's initiative of asking two industrialists (Antoine Riboud, president of BSN, and Jean-Louis Beffa, president of Saint-Gobain)[4] to write reports on the desirable orientations concerning household packaging waste (Riboud, 1991; Beffa, 1991). The government entrusted the industries with the task of formulating what should be done, and widely accepted the proposals made by them. Together with this initiative on packaging waste, a more general reflection on the household waste management reorientation is made by the Ministry of the Environment (and ANRED) within the framework of a broader discussion on the French environmental policy (National Plan for the Environment, 1990). Several new guidelines were proposed to reduce waste production and encourage the development of prevention and valorisation.

The 1992 waste management regime

National objectives and chosen instruments

In 1992, two new pieces of legislation were established. One concerned the entire field of household waste management (July 1992); the other focused on household packaging waste (decree of April 1992). These two pieces of legislation were oriented in the same direction, enabling local authorities to adopt a wide range of alternative techniques for elimination of waste under the best economic conditions. Under the law of July 1992, public authorities ended the former household waste management system based on waste elimination, and were required to support a system built on the objective of waste valorisation. The main objectives were to prevent or reduce the production and noxiousness of waste, to limit its transport and ensure re-use, recycling or other actions to obtain re-usable materials or energy out of waste. After the first of July 2002, landfill sites will only be able to contain ultimate waste (law no 92-646, 13/07/92, article 2.1). All waste categories (except ultimate waste) should be valorised. This new constraint will have to be adopted by all local authorities, who will now have to develop collection and treatment services for household waste to ensure its re-use. The French legislation adopted in 1992 was completed by a special section on packaging waste: industries were compelled to take charge of, or contribute to the elimination of their packaging waste. A target of 75% recycling of household packaging waste must be achieved in 2002.

The new objectives cannot be achieved without a considerable financial effort. To reduce the financial burden, national authorities have decided to set up new instruments. These instruments are used to facilitate or to encourage local authorities to adopt new valorisation schemes, and are established at national level. At local or regional level, subsidies are used in order to reduce the cost of this shift towards valorisation for local authorities. These subsidies vary considerably. In general, the efforts made are important. For example, the *Ile de France* region (Paris and its suburbs) have decided to encourage the implementation of new valorisation schemes (separate collection, incineration with energy recovery, composting, etc.) by investing 115 million Euros in 1994 for a 5-year period. Regional subsidies from *Ile de France* on investments in new equipment cover an important proportion of fixed costs (40% for separate collection and sorting, 25% for incineration with energy recovery and composting). Subsidies concern only investments. Operating costs are always entirely borne by local authorities (except for the case of packaging waste).[5]

National subsidies come from the landfill tax, which was installed according to the law of July 1992. This tax was initially set at 3 Euros/tonne, and re-evaluated in February 1995. It has risen progressively, and attained a rate of 6.1 Euros/tonne on January 1st 1998 (Beguin, 1996). The tax is set according to two goals: correcting a part of the price-gap between landfilling and other techniques, and making a financial transfer (through the setting up of subsidies) to support the investment efforts of local authorities. The returns of this tax supply a fund, which is managed by ADEME:[4] the FMGD (Fund for the Modernization of Waste Management). This organization facilitates the development and construction of new valorisation units. Financial assistance is given in five distinct fields:

(1) Development of innovative techniques in household waste treatment;

(2) Construction of treatment facilities;

(3) Assistance to municipalities and departments;

(4) Elaboration of departmental plans;

(5) Reconditioning of landfill sites and contaminated areas.

The subsidy rates have evolved over time. The rates are currently 30% on separate collection and sorting facilities, 20% on composting and 5% on incineration with energy recovery. These rates will rise in the next months up to 50% for separate collection, sorting facilities, and composting. In 1996, the tax receipts amount to over 110 million Euros. Since its creation, the FMGD has helped 52 large-scale projects, spending around 63.8 million Euros, out of a total investment of 1.3 billion Euros. One can estimate that the tax will raise a total amount of 1.1 billion Euros. It represents a global subsidy of 12–14% on expected investment costs involved by the overall projects, which will be achieved to meet the legal obligations.

Table 3. Sharing out of the FMGD financing for 1996

Type of assistance	Amount (million Euros)	Percentage
Development of innovating techniques	11.4	10.3
Equipment construction	71.4	64.7
Reconditioning of landfill sites	9.2	8.4
Assistance to municipalities	1.5	1.4
Assistance to departments	4.6	4.2
Rehabilitation of polluted sites	12	11
Total	110.1	100

Source: ADEME, 1996

Nevertheless, the total amount of local and national subsidies cannot go beyond certain limits. Investments in new schemes cannot be subsidized beyond 80% (this is an informal rule, not a legal obligation). At national level, it appears that the Regions and the Departments subsidize an average of 40–50% of the investment costs of the main new valorisation schemes. Combined with national subsidies (landfill tax), the total amount of support can rise up to 50–70% for most projects. It shows that financial instruments put into place reduce a very large part of the investment costs borne by municipalities (Table 3).

Eco-Emballages and the promotion of packaging waste recycling

The decree of April 1992 stipulates two possibilities for the fillers. They can either take back and take charge of the treatment of packaging themselves, or contribute to the creation of an authorized organization, which will manage waste instead of the industries. Most of the industrialists chose the second alternative by joining Eco-Emballages.[6] This organization is a private company, the capital of which is mainly held by fillers (70%), but also by distributors (10%) and material producers (20%). The company was authorized by public authorities in November 1992 for a renewable 6-year period. The authorization can be withdrawn if the company does not reach the given targets. The industrialists who subscribe to this company, pay a financial contribution on each packaging and receive in exchange the right to put a green dot on the packaging. The returns of this fee are used to help financing collection and sorting operations managed by local authorities. The fee amounts in average to 0.0015 Euro per packaging (1 centime), and was initially scheduled to rise to 0.0046 Euro in 1998 (3 centimes). This fee equally concerns all types of packaging, whatever the material (Table 4).

Table 4. Packaging fee table (for rigid and hollow packaging)

Volume (cm^3)	Fee per unit (Euros)
>30 0001	0.015
3001–30 000	0.0038
201–3000	0.0015
151–200	0.0007
101–150	0.0004
50–100	0.00015
<50	by weight; ceiling : 0.015

Source: Eco-Emballages, 1992

The green dot contribution level was initially calculated in order to cover the additional cost generated by the valorisation of household packaging waste, i.e. the difference between the separate collection, incineration and sorting cost (for packaging waste), and the collection cost added to incineration with energy recovery (for other household waste). Eco-Emballages financially helps municipalities to install and activate collection and sorting systems for packaging waste, so as to counterbalance the additional costs they have to endure to valorise this waste. This financing is bound by contract. By signing with Eco-Emballages, each local community commits itself to:

(1) Define valorisation objectives, which are compatible with eco-Emballages' task, that is to obtain 75% of valorisation (energy and material).

(2) Establish an action plan and implement a project to inform and aware the population.

(3) Respect sorted material quality criteria according to minimal technical prescriptions.

Local authorities maintain their autonomy of decision, defining the separate collection techniques and dimension facilities according to their objectives. The responsibility of separate collection and sorting of household packaging waste remains in the hands of local authorities. They decide whether to contract with Eco-Emballages, choose the organizational and technical schemes to use and define the relative importance of recycling and energy recovery. Eco-Emballages has no obligation to support all municipalities. The company can refuse to support some of them (in particular those which do not sufficiently promote packaging valorisation, or choose to recycle only one or two packaging materials). This freedom of choice may enable Eco-Emballages to engage the development of packaging valorisation by choosing the timing of action, with a maximum control over its financial commitments (budgetary evolution). Hence, the company is not bound to put into place a large number of separate collection and sorting schemes in a very short period. This may be beneficial because investments in a short period of time could lead to the setting up of inappropriate schemes (those which are available), or difficulties in recycling an increasing amount of sorted waste (due to the lack of facilities).

This staggering of Eco-Emballages' activity can be the starting point of a learning and adjustment process regarding the economic and technical conditions of separate collection, sorting and recycling. This is the sense of the first steps of Eco-Emballages action in 1993. Eco-Emballages decided, before financing any municipal project and generalized its actions, to select 41 pilot operations in which a large range of possible technical and organizational options was tested. The objective was two-fold. First, the running of pilot sites can serve as a basis to calculate the evolution of the financial contributions imposed on fillers, and the support given to municipalities. Second, these experiments enable the local authorities that invest in the development of separate collection and sorting operations to use the gathered information so as to choose the less expensive options and organize their schemes better. The pilot operations have given numerous information, particularly concerning factors which determine separate collection and sorting costs (Cheverry, 1996).

Municipalities commit themselves (for a 6-year period) to respect recovery objectives which are compatible with those of Eco-Emballages. These minimal yields are the result of a double payment of sorted waste: a direct support guaranteed by Eco-Emballages, and a return price guaranteed by material industries responsible for recycling.

Eco-Emballages supports the operation costs of separate collection and sorting schemes of household packaging waste (Figure 1). This is mainly done through direct support to local authorities on the basis of collected tonnage (total tonnage and result per inhabitant).

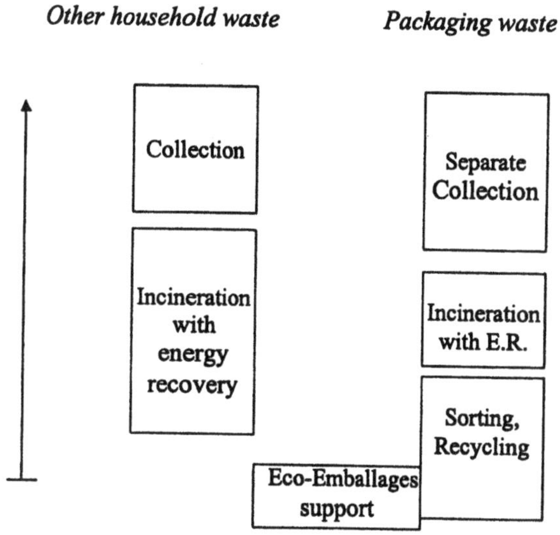

Figure 1. Eco-Emballages' support to local authorities (per tonne of household waste). Source: Gabet *et al.*, 1997

The household packaging waste collected and sorted by local authorities (with the financial help of Eco-Emballages) are taken by material producers at given economic conditions. These producers, organized in five *filières* (each of them in charge of on material), are responsible for the recycling of the waste. They afford guaranteed prices for the return of sorted waste. This is the second part of the financial support given to local authorities. The sorted waste are accepted (and eventually paid) by material producers to local authorities if they respect the minimal technical prescriptions. These prescriptions are based on the material density and the rate of impurity. The financial costs of recycling are borne by the material producers. They have the task to make and to sell recycled products with a secondary material given by local authorities.[7] These two instruments (direct payments and guaranteed return prices) give a minimum payment for each municipality which contracts with Eco-Emballages (Table 5).

Material producers are explicitly incited to play a major role in the running of Eco-Emballages, although they are not considered as responsible for the packaging waste valorisation by the law (the responsibles are the fillers and distributors). They are bound to take back the sorted materials and to recycle them in order to permit Eco-Emballages to attain its targets. In doing so, material producers gain evident interest: they continue to sell their products to fillers at the same economic conditions, and in a growing market (without competition distortion or packaging avoidance). The success of Eco-Emballages is the best insurance for them to avoid a more stringent legislation (Figure 2).

Table 5. Eco-Emballages and material *filières* payments to local authorities (for contracts signed before June 1996)

	Eco-Emballages payment (Euros/tonne)	Material filière guaranteed price (Euros/tonne)	Minimal payment received (Euros/tonne)
Plastic	230	free	230
Aluminium			
– not incinerated	230	155–230	385
– incinerated	75	75–155	150
Iron/steel			
– not incinerated	46	8–30	54
– incinerated	12	0–8	12
Paper/cardboard	115	free	115
Glass	0–8	23	23

Source: Eco-Emballages, 1993a

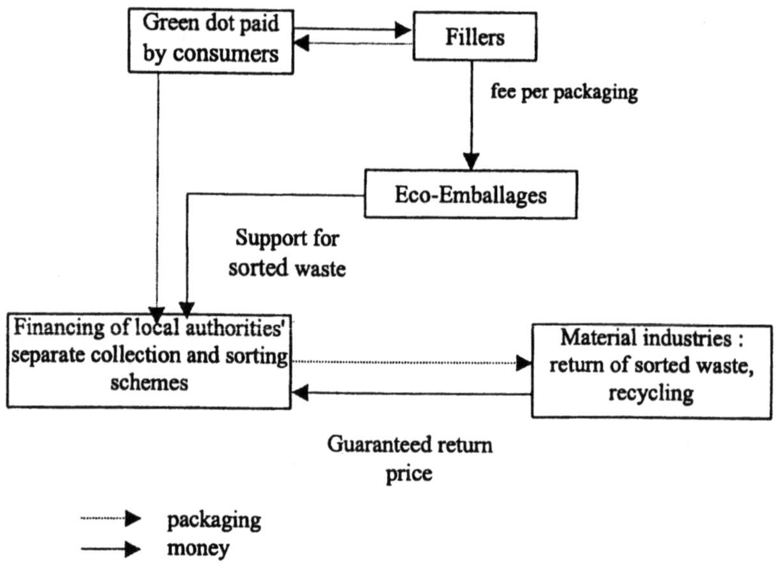

Figure 2. Eco-Emballages: a financial support, an organizational role

Three distingushing principles: decentralization, complementarity and consultation

The French regulatory framework is based on one strong assumption: the implementation of national goals can be achieved in the best economic conditions (for local authorities and consumers) by combining the benefits of three organizational guidelines: decentralization of choices, complementarity of valorisation techniques, consultation in the definition and implementation of goals.

Decentralization of choices

National authorities define the general principles of the French policy, but the means to meet the given objectives stay in the hands of municipalities. municipalities in france are numerous (around 37 000), which is a characteristic almost unique in Europe (Greece also has this feature). The diversity of existing municipalities must be taken into account by public authorities, when the latter establishes objectives and means concerning the household waste management policy. Municipalities are consequently free in their technical and organizational choices (even for packaging waste) and can therefore adapt regulatory objectives to local particularities. In particular, they can decide to join other municipalities to create intermunicipal structures to deal with waste management in a more appropriate level. These structures are widely used in France, especially in rural areas, where municipalities are often too small to finance and manage waste facilities. Chosen objectives (either concerning the limited use of landfill sites, or the valorisation of packaging waste) are implemented on a decentralized basis.

However, this does not mean that municipalities are left to themselves to monitor waste management. Decentralization must nevertheless be coherent with the national household waste management policy. Departmental waste elimination plans are set to ensure this coherence between national requirements and local initiatives concerning the household waste policy.[8] All actors implicated in the departmental household waste policy (the departmental council, municipalities, ADEME, industrialists, associations, Eco-Emballages) co-operate to organize the household waste management, enable compatibility between different initiatives taken by municipalities in this field, and set 5–10 years evolution scenarios.[9] Departmental plans are not only indicative documents. Once defined, adopted by the departmental commission and accepted by a public inquiry commission, the decisions taken at departmental level must be implemented by municipalities. These plans are subject to court supervision and can be cancelled if they do not (or badly) respond to given objectives.[10] Plans are not solely set up by departmental instances. The guidelines are often influenced (if not initiated) by different projects developed independently by municipalities (bottom-up, rather than top-down decision making). This decentralization of choices is set to avoid the economic overcosts which could be created by uniform schemes and organizational structures. The bottom-up decision making could lead to the building of appropriate local schemes, by using local knowledge and information, which cannot be gathered and managed by national authorities.

Complementarity of techniques

Public authorities do not impose precise techniques to be used for household waste management, neither do they indicate on which basis techniques can be combined. On the contrary, financial and economic instruments put into force with the 1992 laws tried to reduce the price differentiation between each valorisation technique. Complementarity of techniques is also a triggering element of the French packaging regime. Both recycling and energy recovery are considered as two valorisation techniques. No hierarchy or ladder has been defined for recycling options. The legislation has not set recycling goals. Eco-Emballages has just indicated in 1992 that 75% valorisation of packaging waste could achieved. This is not a legal obligation, only a preliminary hypothesis. Eco-Emballages has just advanced valorisation targets for each material, but with no obligation to reach them.

Complementarity of techniques may enable a minimization of costs for a given rate of valorisation. This is particularly the case when incineration with energy recovery is integrated into a global treatment scheme, in which sorting, recovery for recycling purpose and composting are incorporated. The search for complementarity may be useful when one looks precisely at the flow of waste, material by material. There are several reasons to expect economic benefits from complementarity, when the techniques used are the most appropriate for a certain type of material. The combined use of separate collection and incineration with energy recovery can, in certain conditions, diminish the overall valorisation cost of household waste, by facilitating the respective running of the two techniques. Hence, separate collection, which concerns in priority materials with low or non-existent calorific value (glass, metal) enables to lighten incineration costs and improve combustion conditions (Cabanes and Oberlé, 1996). Symmetrically, the orientation towards the incineration of certain packaging materials (notably plastic) allows to avoid the costs of material recycling schemes, when this option is too expensive. At a given treatment capacity, using the complementarity of these two options can enable to make significant savings. Separate collection schemes can prolong the life expectancy of an incineration unit, by absorbing part of the increase in the quantity of waste to be treated, and save on fixed cost investments for the construction of a new unit or an additional oven (Bertolini, 1994).[11] Of course, complementarity does not exclude competition between treatment options. Certain materials can be valorised by both *filières*, without one seeming superior to another. This is particularly the case of paper and cardboard.

Consultation

The French household waste system is based on consultation between the implied actors (Figure 3). Industries were associated to the decisional process, which led to the definition of the 1992 legislation. This is particularly true for the packaging waste regulations, for which industries made the first proposal of organization and led all the negotiation process from the beginning to the end. One can see Eco-Emballages as an example of voluntary agreement between public authorities and industrialists. such instruments are generally expected to encourage firm's involvement, to increase their motivation to meet the

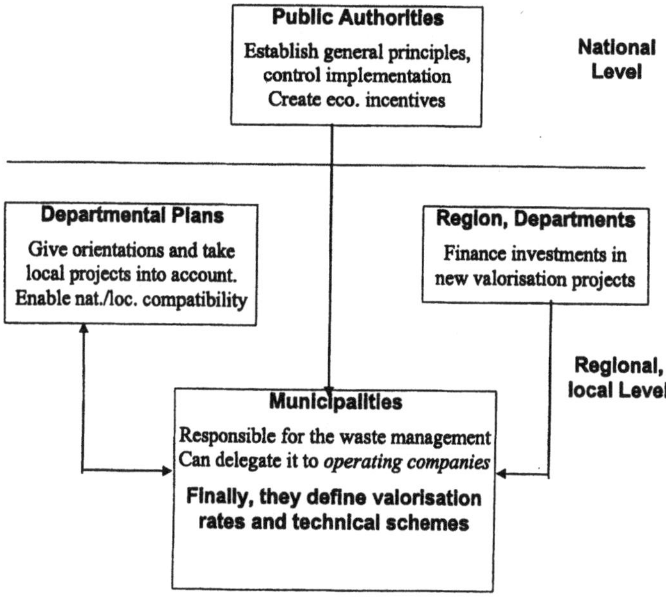

Figure 3. The French household waste regime: a bottom-up approach

valorisation targets and to design an appropriate set of tools (contracts, time-schedule, intermediate goals), adapted to 'real-world' economic and technical conditions. Consultation, and its corollary, flexibility, is also important at local level. The French regulatory framework allows local initiatives, taken at a decentralized level. Municipalities can make different interpretations of the national goals to attain.

The concept of ultimate waste illustrates the room to manoeuvre given to municipalities concerning technical and organizational choices. As previously mentioned, in the year 2002, landfilling will only be limited to ultimate waste. The rate of valorisation to be attained largely depends on the interpretation made on what is considered as ultimate waste. This flexibility enables local actors to carry out frequent adjustments thanks to new information obtained on costs and performances on the one hand, and the valorisation level they want to obtain on the other. The absence of a definition of intermediary objectives between 1992 and 2002 (deadline of the legislation) indicates the intention of leaving room to manoeuvre for actors in the timing of their action. This flexibility may be an economic advantage if the concerned actors use their adaptive capacities to develop a learning process, integrate new information and knowledge, encourage more effective technical and organizational schemes.

The definition of ultimate waste: an ongoing debate

Ultimate waste is the central notion of the French legislation. Its definition may lead to numerous interpretations, and some debates are now occurring between local and national authorities to know what type of interpretations would prevail in the future. In local practice, ultimate waste covers essentially treatment residue (incineration, composting and sorting residue) but cannot be restricted to these categories. Some local authorities consider that ultimate waste can also cover a part of raw waste, which cannot be recycled under current economic and technical conditions. It is the case in deep rural departments where population density is too low to justify the building of additional valorisation facilities (the investment costs borne by each inhabitant appear to be too high). The Ministry of Environment has recently made a more restrictive interpretation of ultimate waste, considering that sorting and composting residue and raw waste cannot be put into the ultimate waste category. The Ministry position is only indicative, there is no obligation for local authorities to follow this position. Debates are occurring at the moment on this question. One can ask oneself whether the ultimate waste definition will progressively sharpen with the improvement of knowledge regarding the current economic and technical conditions, or whether room for interpretation will be left to local authorities.

A first assessment of the new regime

Six years have passed since the setting of the 1992 legislation. One can now have a relatively clear image of the principal outcomes of the general principles put forward by the national authorities in terms of valorisation rates as well as in terms of technical outcomes. Despite rising costs, it seems that the majority of municipalities are starting to engage in the development of valorisation schemes, in which incineration is predominant. New orientations are giving an answer to numerous inadequacies to the current evolution of the French regime.

Positive results but strong shortcomings

Financial outcomes

The 1992 legislation is the starting point of a new and important investment flow. This movement takes place in a context in which costs, and especially operating costs, have risen without interruption for 20 years. Financial implications of the 1992 legislation are important for municipalities, which bear the financial burden of household waste management. Local authorities have to finance the closing of old elimination units and the construction of new ones. They have to equip themselves in separate collection and sorting schemes, in valorisation units (incineration with energy recovery, composting, methanization) or adapt existing plants to the standards' requirements. The total sum invested by 2002 is estimated at 7.7–9.2 billion Euros (BIPE, 1995a). In practice, municipalities do not bear all the investment costs. A part of it is borne by national, regional or departmental authorities, through subsidies (reception of landfill tax, subsidies from Region and Department councils), and operating companies also participate in investment efforts, financing a proportion of the new valorisation projects. The most important part of the municipalities' burden is not the investment costs (largely subsidy), but the operating costs. Two elements are responsible for the operating cost increase. First, the treatment schemes are more expensive to operate, due to the growing stringency of environmental standards (e.g. standards for incineration and landfilling). Second, the technical and organizational schemes to be used to maximize valorisation rates are far more complex than for elimination.

Between 1981 and 1997, the total expenditure (investment and operating costs) of household waste management has practically tripled, going from 1.25 billion of Euros in 1981 to 4 billion of Euros in 1997 (IFEN, 1994 and BIPE, 1997) (Table 6).

Table 6. Evolution of average costs of different techniques (1970–1995)

	1970	1980	1990	1995
Collection	27	34	54	60
Separate collection				180
Controlled landfilling	6	13	31	45
Incineration with ER	24	38	54	80

Costs including operating and investment costs. All costs are constant (reference year: 1997).
Sources: GEERS, 1974; ADEME, IFEN, 1994–1995, 1996

As regards the valorisation of packaging waste, the financial consequences have been relatively limited. Four years (1993–1996) have been spent generalizing the appending of the green dot and the membership of firms to Eco-Emballages. At the end of 1993, only 4025 firms had paid their contribution to Eco-Emballages (30% of the total number of industries affected by the 1992 legislation). Eco-Emballages membership increased to 7120 at the end of 1994, and 8075 at the end of 1995. At the beginning of 1997, almost all the concerned firms have joined Eco-Emballages (approximately 11 000). The green dot now appears on 91% of consumable goods bought by households. By 1997 Eco-Emballages had succeeded in gathering almost all the concerned industries. However, the obligation (by the law) of firms to provide or contribute to the management of their waste for the beginning of 1993 was not respected. To date, only 58 firms had paid a contribution to Eco-Emballages. These firms have been neither controlled, nor sanctioned; consequently, the take-off of Eco-Emballages has been much longer than expected.

The development of the operations of separate collection and sorting increase at the same pace as the payment of contributions. Supports given to municipalities have raised from 3 million Euros (1993) to 67 million Euros (1997) (Figure 4). Consequently, the generalization of packaging waste separate collection and sorting has also been slow. In 1994, only 61 local authorities had signed a contract with Eco-Emballages: this figure rose to 103 in 1995 and 9700 at the end of 1997 (from a total of around 37 000). These contracts should involve approximately 18 million inhabitants, who participate to separate collection and sorting schemes. At present, only around 20% of the French population are involved in packaging waste valorisation (Eco-Emballages, 1998) (Figure 5).

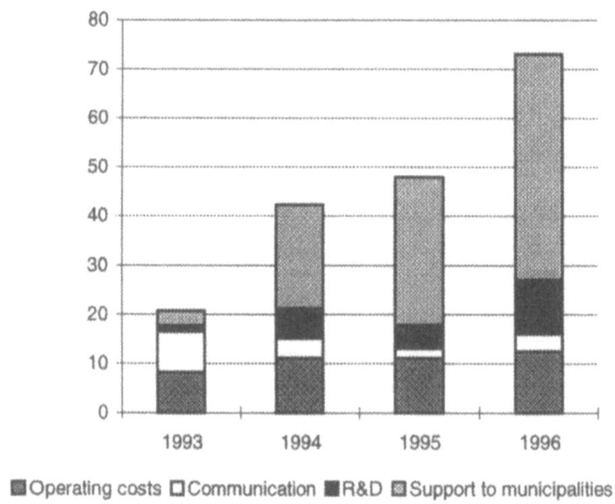

Figure 4. Breakdown of Eco-Emballages expenses (1993–1996) (millions Euros)
Source: Eco-Emballages, 1996, 1997

The objectives: a slow pace of achievement

Despite the financial consequences of the rising costs, municipalities are commiting themselves to the development of separate collection and sorting schemes and installation of valorisation units. The commitments made in the departmental plans (extrapolations drawn from the plans which have been or are going to be published) clearly indicate this tendency. Investments already incurred in the plans (decided by municipal and departmental authorities) amount to 3 billion Euros. This means that between 31% and 38% of expected investments connected with the implementation of the 1992 legislation have

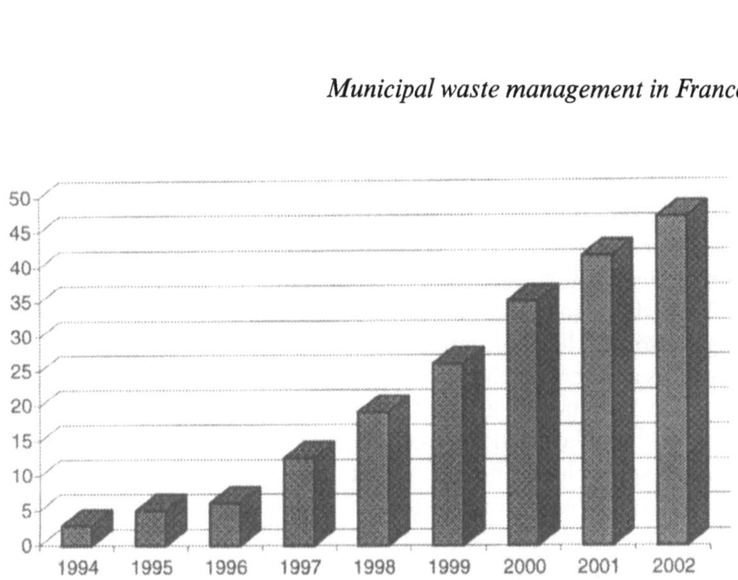

Figure 5. Evolution of population involved in programmes supported by Eco-Emballage (estimations) (million inhabitants)
Source: Eco-Emballages, 1996

already been pledged. Out of that total amount, around 1.38 billion Euros have already been spent (Denby-Wilkes, 1996), which proves that regulatory objectives are starting to be put into practice, and that we are heading towards valorisation (Table 7).

It seems, however, that investments are not sufficient to attain the national valorisation goal by 2002 (that is to restrict elimination to ultimate waste). The number of municipalities that have not yet made the appropriate investments to attain the expected valorisation rates is still significant. The first reason is that the consultation phase before decision making (the discussions between actors in the framework of departmental plans) is much longer than expected. Although these plans had to be established by February 1996, 20% of them had not been by the beginning of 1998 (this figure was almost 50% in July 1997). This curbs the investment flows and the establishment of new facilities.

Table 7. Evolution of household waste treatment techniques (1989–1995)

	1989	1992	1995
Sorting	0.0	0.1	1.7
Incineration	42.3	45.8	46.8
Composting	7.8	7.1	8.1
Landfilling	49.9	47.0	43.4

Source: Bergey, 1997

The other reason explaining this delay is the difficulty that municipalities face in making sufficient investments. Their financial constraints are significant, they are often confronted with real budgetary difficulties and are reluctant to increase local taxes. In addition, the investment costs of the national waste management legislation have been largely underestimated in 1992 (by 10 times). The costs of valorisation schemes rise year after year, which tends to discourage or to curb local initiatives. Local authorities protest against the lack of a reliable assessment of the financial consequences of the national legislation. Some local politicians advocate the need to re-examine the appropriateness of the 1992 regulatory framework (Guellec, 1997).

The municipalities have another reason to make a critical assessment of the national legislation. According to several convergent estimations (ADEME, BIPE),[12] incineration with energy recovery would be the most widely used valorisation technique in 2002.

Incineration would treat between 65% and 70% of the total amount of household waste. This represents a doubling of the quantities of incinerated household waste. The place of other valorisation techniques would be relatively limited. Composting (8%) and recycling (19%) represent together 27% of valorisation, versus 8% for landfilling and incineration without energy recovery (Denby-Wilkes, 1996; BIPE, 1995b). The expected figure concerning recycling is somewhat optimistic. It is far below the current level and it is

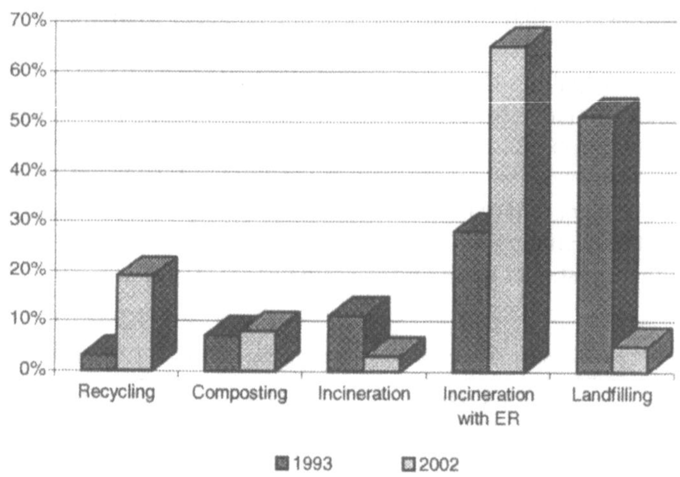

Figure 6. Estimated evolution of the household waste treatment modes (1993–2002) (by treated tonnage)
Source: ADEME, 1996

difficult to see that the 20% will be attained; consequently, the place of incineration may be even more important than expected, and it appears to be predominant for many local authorities. A national legislation which finally aims at establishing 180–200 new incinerators is somewhat difficult to follow without criticism for local authorities which must take into account much opposition to this technique. This is the reason why many local authorities are reluctant to engage themselves in a national initiative which finally would favour the development of incineration, rather than recycling or composting.

The slow pace of achievement of national goals is also seen for packaging waste. In 1995, Eco-Emballages made it possible to re-use 700 000 tonnes of household packaging waste, i.e. 12% of the packaging put on the market (Eco-Emballages, 1996). Of that total amount, 420 000 tonnes were recycled (60%) and 280 000 tonnes incinerated (40%). The material recycling primarily concerns glass, which represents more than 60% of the tonnage of household packaging waste recycled. In 1997 Eco-Emballages treated 1.8 million tonnes of packaging waste (72% recycling, 28% energy re-use). In 1997, the overall

The leading place of incineration

France has the highest number of incineration plants (313) in Europe. More than 8 million tonnes of household waste are burnt each year, which corresponds to the most significant flow of household waste treated by incineration in Europe. In 2002, estimations assert that the number of incinerators, according to the consequences induced by the 1992 legislation, should stay at a high level (310–350), and 180–200 new plants should be created (BIPE, 1995a). These plants would be able to treat almost 16 million tonnes of waste, representing 65% of the total amount of household waste. The types of incinerators used are relatively diverse. Most incineration plants (in number, not in treatment capacity) are composed of small-scale and old units (72%). The rest consist of huge plants, treating mostly household waste. These plants generally provide energy recovery. French incinerators ought to respect European emissions standards set in 1991, but estimations made in 1995 show that only 44% of the incineration plants respect these standards, mainly the biggest ones. Two main trends can be underlined. First, all new facilities provide energy recovery. Incineration facilities without energy recovery are to be closed down or modernized. Second, the average scale of incineration plants tends to increase. This is due to two elements: on the one hand, the tightening of emission standards requires installation of expensive depollution systems, which cannot be redeemed for small-scale facilities. On the other hand, municipalities tend to benefit from economies of scale provided by large incineration plants.

rate of recycling and incineration of household packaging waste reached 31%, with a recycling rate of 22%. The increase of packaging waste recycling between 1995 and 1997 was almost exclusively due to glass recycling, which benefited from a long and successful experience prior to the starting-up of Eco-Emballages programme.

Table 8. Quantities of recycled materials (1995–1997) (tonnes)

Material	1995	1996	1997	Total*	% recycled in 1997
Steel	127 000	146 000	175 000	500 000	35
Aluminium	1 200	1 100	n.a.	30 000	n.a.
Paper/card.	24 000	45 000	75 000	1 800 000	4.1
Plastics	120 000	19 000	31 000	1 200 000	2.5
Glass	256 000	370 000	1 062 000	2 300 000	46
Total	420 200	591 000	1 300 000	5.83 million	22

*Total quantity of household packaging put on the market

Source: Eco-Emballages, 1996–8

Correcting evolution of the regime

Areas of dissatisfaction

The new regime defined in 1992 has already been challenged in two main fields: the place given to the composting option and the efficiency of the organization required to develop and promote packaging waste recycling.

A neglected option: composting

Composting represents a diminishing part of household waste treated: the percentage of household waste that is composted has gradually decreased during the last 15 years, from 10% in 1981 to 8% in 1997. In 1993, 1.48 million tonnes were treated, compared with 1.34 million tonnes in 1989. This slight progression is due to the increase in the mass of waste produced. However, this must not mask the stagnation of this treatment mode. The number of composting units has fallen from 77 units in 1989 to 68 in 1993, and the compost produced at the moment amounts to only 600 000 tonnes/year. At the beginning of 1998, only 30 operations of co-composting (household waste and organic waste from parks and individual gardens) existed in the rural areas of the country, compared with 100 in Germany and 400 in Austria (Besançon, 1998). The same is true for very small composting structures (less than 200 tonnes/year). There exists only one site in France compared with more than 200 in Sweden. The situation is therefore far from being satisfying, and environmental associations heavily criticize it.

Composting requires a market for the compost produced. in order to minimize transport costs, compost must be made on land which is as near as possible to the composting unit. Farmers are, however, less and less inclined to buy this product because urban compost still has a bad reputation. In the past it was composed of unsorted waste, and filled with glass and plastic. Legal proceedings were taken against the use in agriculture of this type of compost, because of accidents such as the death of cattle coming from the ingestion of plastics, or wounds caused by the pieces of glass in the compost. Although efforts have been made by some local authorities to ensure a high quality compost, this bad reputation is still justified by the predominance of poor quality compost on the market. Facing this quality uncertainty, many farmers refuse to spread compost on their land, whether the compost is of good or bad quality; as a result, the compost demand is low.[13] The creation of a new quality standard has not been sufficient to secure and develop the compost market.

Furthermore, the compost market suffers from the competition of other chemical fertilizers. The agricultural co-operatives that receive part of the profit of sold fertilizers do not consider the compost market well. Faced with the fertilizer lobby, compost promoters do not carry much weight. The fact that the compost market is not competitive can also partly be explained by the lack of incentives taken to develop composting techniques. Finally, the expansion of composting is blocked by a restrictive and complex legislation, which could cause local authorities not to invest in such a system. Composting is regulated by different laws, concerning spreading conditions on land which tend to become more and more strict. Many social and economic actors have asked the government to take the appropriate measures in order to restore trust in the quality of the obtainable compost and to orientate their financial support in favour of this treatment technique.

Eco-Emballages and the promotion of packaging waste recycling

Criticism has been made of Eco-Emballages and its willingness to support local authorities' efforts to develop separate collection, sorting and recycling schemes. Critical points are focused on the lack of supports given to municipalities, the firms' weak commitments, and the obstacles to the development of markets for recycled materials.

The first available economic assessments show that the municipalities engaged in separate collection and sorting do not always benefit from sufficient support from Eco-Emballages. At first glance, this result may sound strange. According to Eco-Emballages, support given to municipalities ensures that they can recoup the additional costs of the separate collection and sorting schemes for packaging waste. The problem is that Eco-Emballages' calculation is made on a national basis, for an average flow of packaging per inhabitant and with high levels of separate collection and sorting rates. In practice, this national-based average additional cost and the average amount of support given corresponds to very few real local situations.

In some cases, the Eco-Emballages supports are sufficient: this is the case of municipalities which are characterized by large population (hundred thousand inhabitants), strong involvement in incineration with energy recovery and well-organized collection schemes with a high frequency (5–7 collections per week). These cities bear relatively low additional costs when they implement separate collection and sorting schemes for packaging waste, because they can easily substitute one of their collection tours for a separate collection without re-organizing all the collection system. Optimisation of the joint use of traditional collection tours and separate collection is therefore possible with full use of existing facilities.

Nevertheless, a large number of urban municipalities are confronted with insufficient support from Eco-emballages, mainly because they cannot combine traditional collection and separate collection without significant additional costs. The lower collection frequency does not allow separate collections to be substituted for traditional collection tours. Municipalities may also have some difficulty in reaching high collection rates per inhabitant and per material. This reduces the amounts paid by Eco-Emballages (especially for plastics). This is the case of Créteil, a medium-size urban municipality (88 000 inhabitants), a town which suffers from a deficit in its financial balance of its packaging waste separate collection and sorting activities (after taking into account Eco-Emballages' support). It appears that one of the most important element explaining this deficit is the low collection and sorting rate of plastics (1.5 kg/inhabitant/year), far below the level required (5 kg/inhabitant/year) to benefit from the maximum support of Eco-Emballages (Table 9).

Table 9. An example of Eco-Emballages' insufficient supports: Créteil

Number of inhabitants	88 000
Packaging waste collected	44 kg/inhabitant/year
Separate collection and sorting costs	207 Euros/tonne
Traditional collection and incineration	122 Euros/tonne
Additional costs of packaging waste valorisation	85 Euros/tonne
Support given by Eco-Emballages	52.7 Euros/tonne
Financial deficit per tonne	32.3 Euros
Total financial deficit (1996)	375 000 Euros
Total financial deficit (1994–96)	700 000 Euros

1996 figures.

Source: City of Créteil, 1997

The difficulties are even worse in rural municipalities, where the implementation of a separate collection scheme in addition to existing collection schemes induces high additional costs. In addition, the packaging flow is generally below the urban level since each inhabitant produces a lower quantity of packaging waste. This implies that the Eco-Emballages rates of separate collection are even more difficult to attain.

This situation can explain why municipalities are not committing themselves to a large and ambitious development of separate collection and sorting schemes. They face a lot of uncertainties, which in the long run could curb their willingness to promote valorisation. If they do not have a sufficient packaging waste flows and do not attain the Eco-Emballages requirements regarding separate collection rate and quality of sorted materials,[14] they could be in position to bear recurrent financial losses.

Eco-Emballages is not responsible for the technical and financial results obtained by separate collection and sorting schemes. It is the municipalities' task to obtain sufficient performances to be reimbursed for the additional costs supported. They bear all the risks and all the responsibilities regarding packaging valorisation and industry's involvement is reduced to a minimal role. In contractual terms, one can say that the municipalities which engage themselves into separate collection operations are those which are risk-neutral (or indifferent to risk-bearing), or do not have any a priori information about the potential consequences of their involvement. Those which are risk-adverse will not be inclined to contract with Eco-Emballages and therefore will not be in position to adopt separate collection schemes. Once engaged, the risk-neutral municipalities will gather information and may have some reasons to reduce the scope of their separate collection and sorting operations, or even to stop it when social pressure towards valorisation will slow down. The asymmetry of contractual terms between Eco-Emballages and municipalities may be a strong rationale to curb the adoption of recycling schemes.

If municipalities are not always sufficiently supported, it also appears that industrialists are not inclined to work towards the rapid development of recycling. Two reasons explain this. First, the more recycling develops, the more Eco-Emballages would have to take on financing, and the higher the contributions paid by the fillers. Second, the more recycling develops, the more material producers would be obliged to take back large quantities of waste materials, and the more they would have to confront technical and financial problems, especially for plastics. Eco-Emballages may prefer to curb the promotion of recycling and to let energy recovery take the leading place in packaging valorisation.

The company has several means to do so. Considering the fixing of minimal technical prescriptions and Eco-Emballages' freedom to establish contracts with local authorities, industrialists have the means to limit the financial needs expressed by the municipalities. Referring to contractual terms, one can say that the incentive structures of contractual relationship are not well aligned with the legal obligations to be attained. Eco-Emballages is not positively incited to promote recycling and has several means to limit financial supports given to separate collection and sorting schemes. If the terms of the contractual relationship between Eco-Emballages and the municipalities are not re-examined, the packaging recycling rate is likely to stay at a permanently low level. This re-examination requires that industrialists accept to involve themselves beyond the financial limits they have fixed. Paradoxically, the importance of consultation in the definition of the 1992 legislation seems to explain a major part of the existing situation. Industrialists played a leading role in the formulation and the implementation of this legislation. Eco-Emballages is the product of a long negotiation process, which was held to persuade all the industrialists (with possible conflicting interests) to enter in a collective action to promote household packaging waste valorisation. The stake was in particular to fill the gap between fillers and material producers. This consensus-driven process has led to important drawbacks.

First, it has been necessarily to avoid any additional cost disturbances between material *filières* (the fees paid by fillers are not differentiated: the charge is equal whatever the material). This allows, on the one hand, avoidance of new market distortions between material producers and, on the other hand, to allow fillers to continue to use the whole range of packaging materials without any restriction. Second, industrialists have tried to minimize the financial burden they ought to bear (using the additional cost calculation, and externalizing responsibilities, performances, uncertainties and risks towards municipalities). Third, they have tried to gain a maximum control over their involvement in collection actions. For that reason, they set a limit on their financial commitment for recycling promotion at the very beginning of the process (1991).

This does not mean that any industrialists' involvement in the implementation of a legislation leads to a minimal programme erasing a large part of the incentive schemes which are necessary to reach a high recycling level. The German experience shows an opposite point of view. The French negotiating process for the setting of Eco-Emballages and its outcome results from the level of constraint put on industrialists by national authorities. The public authorities' requirement regarding packaging valorisation can be considered as weak (75% valorisation for 2002, no intermediate targets, no recycling goal to be attained, a responsibility leading to the bearing of additional costs and not the entire costs of packaging valorisation, etc.). With such lack of pressure, it is not surprising that industrialists have found an appropriate ground to define a minimum programme, based on weak involvement.

In fact, this low level of public requirement regarding packaging waste valorisation is the consequence of the decentralization of choices and complementarity of techniques. These two elements have important repercussions on the setting of national requirements and on the driving of negotiation processes. Decentralization of choices leads to put the responsibility of packaging waste treatment in the hands of municipalities. Therefore, they can only be supported in terms of additional costs and not on a full-cost basis.

Weak involvement of industrialists follows. Complementarity of techniques makes sense only if there are no specific targets set for the different packaging valorisation techniques. However, this enables industrialists to limit their support to separate collection and sorting schemes and to give the leading role to energy recovery. Decentralization, complementarity and consultation may have some important drawbacks when the former two diminish public authorities' requirements and badly influence the consultation stage.

The difficult emergence of material *filières*: the case of plastics

In the framework of the Eco-Emballages system, plastic packaging stands as the most challenging issue in terms of material recycling. In 1997, plastic represented less than 2.5% of the total amount of materials recycled. Valorplast, which is in charge of the matter, has to face a situation very different according to the type of plastic (PVC, PET, PEHD). Until the last two years, France was the most involved country in the use of PVC, in particular regarding bottles. There was a contrast with the existing situation in other countries where PET had already definitely supplanted PVC. This situation created a sort of bottleneck for the recycling output of the latter, whereas France was over-equipped with PVC regeneration units compared with those for PET. Second, the technical prescriptions on the quality of sorting imposed to local authorities are much less constraining than standards adopted by the european market. This fact was aggravated by the still higher rate of PVC bottles in France, which made it more complicated to obtain good quality sorting. In order to find the needed recyclers, private operators have to proceed to a second sorting phase. Consequently, the cost of the PET collected in France is too high to be competitive. This is clearly not an incentive to develop plastic recycling. The progressive disappearing of PVC bottles should help the situation to be more favourable in the end to the sorting of plastic bottles.

The third problem lies on the economic balance of the network with the recycling recent price evolution of the virgin material, which plays against an increase in the use of regenerated PET. While the price of the virgin material was at 1.5–1.8 Euros/kg, in January 1997 the market level dropped to 0.4–0.5 Euros/kg. At such a price level, regenerated PET is not in a position to compete with the virgin material. On the one hand, recycled PET offers a poorer quality and thus has more limited outlets,[15] on the other, recycling the materials generates various types of costs among which transformation costs (sorting out, shredding, washing, drying) and transport costs. With the drop in the price of the virgin material, which has taken place over the past months, the economic profitability of the PET recycling *filière* is questioned. The future of the PET *filière* is conditioned by the existence of additional financing sources that will have to be found within the system if the PET network is to be maintained. Accordingly, the creation of some valorisation *filières* relies less on technical consideration than on price-related aspects. The development of mechanical recycling for plastic packaging is conditional upon the cost difference between virgin and regenerated resin, which is itself tightly linked to the price fluctuation of oil. Today, the balance of the system implemented within the framework of Eco-Emballages is currently the subject of increasing pressure.

> At present, both collection activities and sorting out operations are supported financially by Eco-Emballages, but not the next stages, which concern the regeneration and recycling of the product. In the current situation, the mere survival of some material networks appears to depend on the ability of financial structures to ensure the economic balance of all the filières, from collection activities to the actual recycling of the product. In this context, the further development of packaging recycling could rest on the adaptation of Eco-Emballages' financing system to the reality of the market. In the future, Eco-Emballages' financing system could expand its scope of intervention to ensure the economic viability of the material *filières*.

Initiatives towards a new equilibrium

The main orientations given to the French household waste management in 1992 have driven the national regime in quite an unsatisfactory situation concerning its capacity to promote the development of recycling. The initial organization of Eco-Emballages favoured too much incineration with energy recovery, compared to the announced objectives. The observation of departmental plans is, as already stated, a good indicator of such a bias. The average of planned incineration is of about 65–70% of household waste treated. New and important modifications were decided in the 1996–1998 period.

Firstly, Eco-Emballages decided in June 1996 to redefine the basis of its supports to local authorities, in order to foster the development of recycling. For paper and plastic, Eco-Emballages' payments increased according to the collection output obtained by the municipality: support for plastic amounts to 230 Euros/tonne if the annual output per inhabitant is less than or equal to 2.5 kg, rising to 692 Euros/tonne if the output reaches 5 kg/year or more (Perrier, 1996). As collection and sorting become more efficient, Eco-Emballages' support also becomes more important. Support for energy recovery is lower than support for separate collection and sorting. This support is only obtained if incineration is complementary to recycling, and depends on the amounts of waste recycled. Additional support is made to communication plans and to the setting up of separate collection and sorting schemes (temporary support) (Table 10).

Other initiatives or political orientations have been made. This is particularly true for departmental plans which give too much importance to incineration. In 1997, several declarations of the Minister of Environment attempted to make local authorities understand, mostly at the departmental level, that the objectives included in the plans were unsatisfactory and that changes of orientation ought to be made. An official note from the Ministry of Environment (28/04/1998) clarifies the notion of ultimate waste which, until now, was assimilated by many actors to the output of incineration. The official note specifies that this is 'the non-recoverable fraction of waste and not only the

Table 10. Eco-Emballages direct payments (for contracts signed after June 1996)

Collected material to be recycled	Euros/tonne
Glass	
Collective containers	3.07
Separate collection	7.7
Plastics	
<2.5 kg/inhabitant/year	230
2.5 to 5 kg/inhabitant/year	230
>5 kg/inhabitant/year	692
Steel	
Separate collection	46.1
After burning	11.5
Aluminium	
Separate collection	230
After burning	77
Paper/cardboard	
<5 kg/inhabitant/year	115.3
5–15 kg/inhabitant/year	115.3–254
>15 kg/inhabitant/year	254
Collected materials to be burnt*	
0–25% of recycling	15.3
25–50%	15.3–77
up to 50%	77

*For 1 tonne recycled, 1 tonne incinerated is supported (except for plastics 1 to 3). Only if all 5 materials are recycled and if minimal limits for plastics and paper/cardboard collection are attained (2.5 kg/inhabitant/year for the former, 1 kg for the latter).

Source: Eco-Emballages, 1996

output of incineration'. This point is important as it gives less the impression to local authorities that incineration cannot be avoided and that in some cases, landfilling may be an acceptable solution for waste being discarded from sorting or composting activities.[16]

The note also insists on the greater emphasis to be given to recycling (including composting) with a long-term objective of at least 50% (this objective is indicative, not legal). This objective is valid at a national level but ought not to be uniform, in order to take territorial particularities into account. Nevertheless, some departmental plans, as the one of the Yvelines, which schedules 85% of incineration, will have to be strongly revised. The official note is also important from another point of view. It tries to give a more operational interpretation of the principle of proximity. Mainly because of the lack of flexibility of the incineration capacities, and of the required economies of scale which would incite departments to maintain a large room for this technique, the geographical limits of departments are often felt as too restricted. The note calls for co-operation between departments on this point in order to obtain environmental and economical benefits.

These two points put together (the modification of payments of Eco-Emballages to municipalities and the correction of departmental plans in order to orient their objective as a whole, towards at least 50% of recycling) give a new impulsion to the national policy of household waste management. One must note, however, that the tripling (from 0.0014 Euros per package to 0.0042 Euros per package) of contributions paid by firms to Eco-Emballages, originally scheduled for 1998, has not been applied because of the sum obtained with the former level of contribution not being already totally used. The situation was clearly due to the insufficient financial incentives for local authorities to develop sorting and composting. Together with the absence of strong legal incentives, it has driven to a low effort of investment in sorting and a report of the required investments on incineration with energy recovery as an easier solution. The official note of April 1998 was greatly expected by municipalities and departments and, according to E. Guillon, director of Eco-Emballages, immediately produces an increase of new contracts between Eco-Emballages and municipalities in order to promote sorting. The official note also modifies the approach to the management of household waste and other municipal waste. The previous objective of recycling focused on the packaging fraction of waste, while now it is the whole flow of household waste which is concerned. It is a clear indication that composting, which has been neglected up to now, will gain importance in the plans of municipalities and departments.

Conclusion

France has adopted an original approach concerning household waste management. The initiative implemented in 1992 was based on consultation, decentralization and complementarity of techniques. Firms participated in the definition of regulations, notably concerning household packaging waste, and local actors were involved in the definition of departmental plans. Municipalities became responsible for their technical and organizational choices, and were able to find appropriate means to combine regulatory requirements with local peculiarities. The different actors involved showed a concern in reaching objectives at least cost.

Nevertheless, the French options have important drawbacks which, in the eyes of public authorities and of most of the implied actors, impede the realization of their valorisation objectives. A structural bias in favour of incineration has been emphasized. This bias is mainly due to institutional factors, including weakness of the public and private instruments which should promote the development of recycling and composting. Up to the end of 1997, municipalities adopted technical schemes in which incineration played a leading role. A regime willing to promote a complementary use of valorisation techniques led to the development of a technological trajectory based on one prevailing technique. It could seem paradoxical that this argument on the complementarity of techniques has been introduced mainly to leave a place for incineration which was threatened by European directives. It allowed the pursuit of the former tendency of the French household waste policy, which was mostly favourable to this technique, also because of the inadequacy of instruments used to promote alternative technologies. In a decentralized context, if incentive structures are not well suited, then the logic of technical adoption shifts towards incineration.

The French regime seems to be able to make significant changes and to incorporate adjustment procedures. With the official note of April 1998, for the first time, public authorities give a real limit to the expansion of incineration. Nevertheless it remains only a political orientation, while nothing has changed from the point of view of financial incentives, which appears to be decisive in orientating the behaviour of actors implied. Although significant changes were already introduced by 1996 with Eco-Emballages, increasing the level of financial support for some materials and making recycling more attractive, they appeared to be insufficient to address all current difficulties and the general disequilibrium between financial flows and physical targets. No mechanism has been introduced in order to promote composting, which was, until the end of 1997, almost ignored by the regulator and which now is considered as a priority. The next step for correcting the observed drawbacks, could be the introduction of new financial incentives aiming at pushing local actors to go in the direction required by the regulator and in line with European policy.

Notes

1. This text has been produced with the help of Y. Serret and O. Godard.
2. Household waste, 20.5 million tonnes. Other municipal waste, 4.5 million tonnes (sewage sludge), 3 million, city-cleaning and waste from parks, 1 million, others 500 000 tonnes).
3. Valorisation is a term that covers recycling, composting, incineration with energy recovery and any other form of energy recovery.
4. BSN has now changed into Danone, which is a leading company involved in food products. Saint Gobain is a glass producer.
5. These subsidies are not given to any new project. The project needs to respect certain conditions (high valorisation rate, integration in a global scheme, etc.). ADEME is a governmental agency which works on environmental and energy-related themes, helps municipalities in their choices through advice given and available publications.
6. Two other societies have been authhorized by the government: Adelphe is in charge of the recycling glass bottles for wine and spirits, and Cyclamad is responsible for that of drug packaging.
7. The price difference between the return of sorted materials and the making and selling of recycled goods (or goods which incorporate a part of secondary materials) is borne by material producers. This difference may be positive or negative. It depends on the raw material spot market on one hand, and on the price of additional sorting and regeneration of sorted materials, on the other hand. The possible profit is not shared with municipalities.
8. The term Department is used to define the French notion of *Département*, which has no equivalent in English language. The *Département* is an administrative territorial cutting of France (99 in all), which is bigger than a commune and smaller than a region (22 in all).

9. Departmental waste elimination plans put into practice the proximity principle. Waste produced in a department must preferably be treated in the same zone. This responds to a double requirement: the amelioration of the degree of social acceptability of treatment projects, and the solution to waste imports coming notably from Germany.
10. Up to the beginning of 1998, only one plan was cancelled, because objectives were not precise enough with regard to the requirements of the decree of 1993.
11. As an example, in the region of the Nord-Pas-de-Calais, the running of six sorting centres avoided the construction of a new incineration plant (of a capacity of around 80 000 tonnes), which would have added to the 16 already existing units.
12. These estimates are based on departmental plans published mid-1997, which represent a population equivalent to 35–40% of the French total population.
13. In a 1990 inquiry upon composting plants, the total amount of money generated by the selling of compost was very low, between 0% and 10% of the global composting cost of each plant.
14. Quality prescriptions for sorted materials are defined by industries (in consultation with local authorities' representatives) which take back materials. This may be seen as a paradoxical situation: local authorities are obliged to supply standardized raw materials to the concerned industries, to be ensured that the latter will take care of their waste. If standards are not respected, industries can withdraw. The fixing of minimal technical prescriptions therefore seems to be a critical operation for allocating responsibilities and risks relative to waste management.
15. The possibility to use recycled PET for the production of food containers has been given in December 1997 by the French sanitary authorities. It should help PET to find more outlets than before.
16. Nevertheless, the articulation of the French regime with the coming directive on landfilling has not been clarified.
17. These figures representing an average of the contributions which are calculated according to the volume of the package.

References

ADEME, La taxe sur le traitement, le stockage des déchets et les aides financières correspondantes, Angers, 1996.

ANRED, Etat des réalisations des usines d'incinération d'ordures ménagères avec récupération d'énergie, Paris, 1987.

ANRED, Les décharges contrôlées, Angers, 1989.

ANRED, Les Ordures ménagères, Observatoire National des Déchets, 1991

Beffa J.L., Rapport sur la valorisation des déchets d'emballages domestiques, rapport pour le Ministre de l'Environnement, 1991

Beguin D., Bilan de trois ans de soutien financier à la modernisation, in ADEME (ed.), Déchets municipaux. La modernisation en marche, Angers, 1996.

Bergey J.-L., Les évolutions dans les déchets: conséquences techniques et financières, TSM, n°12, 1997.

Bertolini G., 1994, Economies sur l'incinération d'ordures ménagères résultant d'une réduction à la source ou de collectes sélectives (notamment de fractions sèches et en particulier d'emballages), T.S.M., vol. 89.

Besançon A., intervention au colloque Composter les déchets. Les raisons pour agir, Amiens, avril 1998.

BIPE, Perspectives d'évolution de la gestion des déchets ménagers et assimilés, Paris, 1995a.

BIPE, Prospective des déchets industriels et ménagers en France à l'horizon 2010, Paris, 1995b.

BIPE, Le marché des activités liées aux déchets. Situation 1996 – Perspective 1997, ADEME Editions, Paris, 1997.

Cabanes A. and S. Oberlé, 1996, Valorisation énergétique des ordures ménagères par incinération. Etats des lieux et projets, Déchets. Sciences et Techniques, no. 4.

Cheverry M., Les références communes et la maîtrise des coûts, Conférence Maîtriser un programme de collecte sélective, Ademe et Eco-Emballages, Lyon, 22 octobre 1996.

Defeuilley C. and O. Godard, 1997, The economics of recycling in France: institutional framework and technological adoption, International Journal of Environment and Pollution, 7.

Defeuilley C. and S. Lupton, The future place of recycling in household waste policy: the case of France, Resources, Conservation and Recycling, 1998, vol. 24.

Denby-Wilkes V., Politique nationale: sa traduction en termes d'équipements et d'investissements, Conférence La gestion des déchets municipaux: comment mieux gérer au moindre coût, Paris, 26 juin 1996.

Eco-Emballages, Demande d'agrément, Levallois-Perret, 1992.

Eco-Emballages, Eco-Emballages et les collectivités locales, Levallois-Perret, 1993a.

Eco-Emballages, Bilan de l'appel à candidature aux opérations pilotes d'Eco-Emballages, Levallois-Perret, 1993b.

Eco-Emballages, Annual report, Levallois-Perret, 1996, 1997 and 1998

Gabet A., R. Lantreibecq, J. Vigneron and L. Guerineau, 1997, La collecte sélective des déchets ménagers. Triselec: la bonne affaire, Economica, Paris, 1997.

GEERS, Les déchets solides, propositions pour une politique, La documentation française, Paris, 1974.

Gellec A., 1997, Déchets ménagers: pour un retour à la raison, Assemblée Nationale, Commission de la production, Rapport d'information no. 3380.

Godard O., The new French packaging waste recycling mechanism. Looking for appropriate incentives, International workshop on the Institutional Aspects of Economic Instruments for Environmental Policy, Copenhagen, May 20–21 1997.

IFEN, En dix ans, augmentation de 80% en francs constants, Les données de l'environment, no. 7, 1994.

IFEN, Les Comptes de la dépense de protection de l'environment. La gestion des déchets 1990–1993, Coll. Etudes et Traveux, no. 9, Orléans, 1996.

Perrier M., Les financements au fonctionnement des collectes sélectives, Conférence Maîtriser un programme de collecte sélective, ADEME et Eco-Emballages, Lyon, 22 octobre 1996.

Quirion P., La gestion des déchets d'emballages ménagers en France et en Allemagne: éléments d'évaluation économique, mémoire de DEA, Paris I, 1994.

Ribaud, A., 1991, Emballage et environment, Mission Eco-Emballages: Rapport final, Annexes, Vol. 1, Tome 2, p. 305–309.

Theys J. and L. Chabason, 1990, Plan national pour l'Environnement, Ministère de l'Environnement, Paris, 1990.

Whiston T. and M. Glachant, 1996, Voluntary agreements between industry and government. The case of recycling regulations, in Lévêque F. (ed.), Environmental Policy in Europe. Industry, Competition and the Policy Process, Cheltenham: Edward Elgar.

CHAPTER 4

Municipal waste management in Italy

Paolo Bertossi, Antonio Kaulard and Antonio Massarutto

Introduction[1]

This chapter provides an outline of the Italian regime for managing municipal solid waste (MSW). The waste industry in Italy is undergoing a rapid process of development and modernization. This process resembles that occurring in other EU countries, in the sense that MSW collection and disposal services are evolving from essentially end-of- pipe activities, which are concerned with the environmental impact of disposal, but constrained by given quantity and quality of waste flows, towards a far more complex model, where the objectives are waste avoidance, reuse, and valorization, according to the ladder principle, as well as the principles of self-sufficiency, proximity and polluter-pays, in accordance with Dir 91/156/EEC. Waste disposal operators are no longer the only relevant actors, since the focus of waste policies has gradually shifted towards materials and commodities. MSW policies are no longer local health policies, aimed at eliminating waste from streets and households, and have entered the domain of environmental policies, focused on the allocation of scarce environmental and territorial resources.

What makes the Italian case special is perhaps the rapidity of the transformation and the somewhat chaotic evolutionary process. What other countries have achieved in 30 years requires a much faster evolution in Italy, given the delay accumulated in the past. The capacity of environmental policymakers to govern the evolution is much weaker, while the policy network is dominated, at the local level, by waste facility operators. Therefore, in order to understand the Italian regime, we feel it is important to focus attention not only on the formal regulation procedures, but also, and even more so, on the informal practices which underlie the action network.

At the beginning of 1997, a new law (Dlgs 22/97) was passed in order to comply with the EU Directives implementing the Environmental Action Program V. The new law has set up a completely new regime, already anticipated by many local municipalities, and is likely to accelerate the process

N. Buclet and O. Godard (eds.), Municipal Waste Management in Europe, 121–169.

of change. However the constraints described above not only remain valid, but, are also likely to increase further in importance, since the new regime has been set up without a complete and satisfactory achievement of the objectives set by the previous regime.

The chapter is organized as follows. In section 2, we analyze the historical development of the regime; we argue that the present phase lies somewhere in the middle between two regimes in equilibrium, which we have called baseline (effective removal of waste under the external constraints arising from environmental regulation) and green target (sustainable management of materials flow through the economy).

In section 3, the analysis is concentrated on the present transition phase. Data and information about the structure of the Italian waste industry are provided, together with a description of the financial structure of the MSW industry and case-studies of management schemes.

Section 4 is concentrated on the packaging issue, analyzing the experience of Obligatory Syndicates (1988–1997) and the emerging of the new regime after Dlgs 22/97.

In section 5 we conclude by the reviewing the prospects for a proper implementation of the EU MSW policy framework in Italy, focusing attention on three main areas: the effective implementation of environmental regulation, the difficulty in achieving territorial integration or coordination of efforts, and the various combinations of local policies and emerging trends in the MSW industry.

Historical outlook: the development of the Italian household waste management regime

Overview

MSW management became a policy issue early on, together with industrial and urban development. Leaving aside the north-west of the country, this coincided approximately with the economic boom of the years immediately following World War II. During the past 50 years, MSW policies have undergone a significant transformation from a very local issue, mostly related to health and urban landscape policies, to an issue underlying national environmental policy. In this section we try to indicate the milestones that mark this development path.

As shown in Figure 1, we can identify 4 significant phases, each of which can be characterized by a different set of problems and priorities, a different management system, a different regulatory approach, a different division of labour and different pattern of market segmentation.

We turn now to a brief outline of each phase in order to be able to later on identify the main features of the current regime.

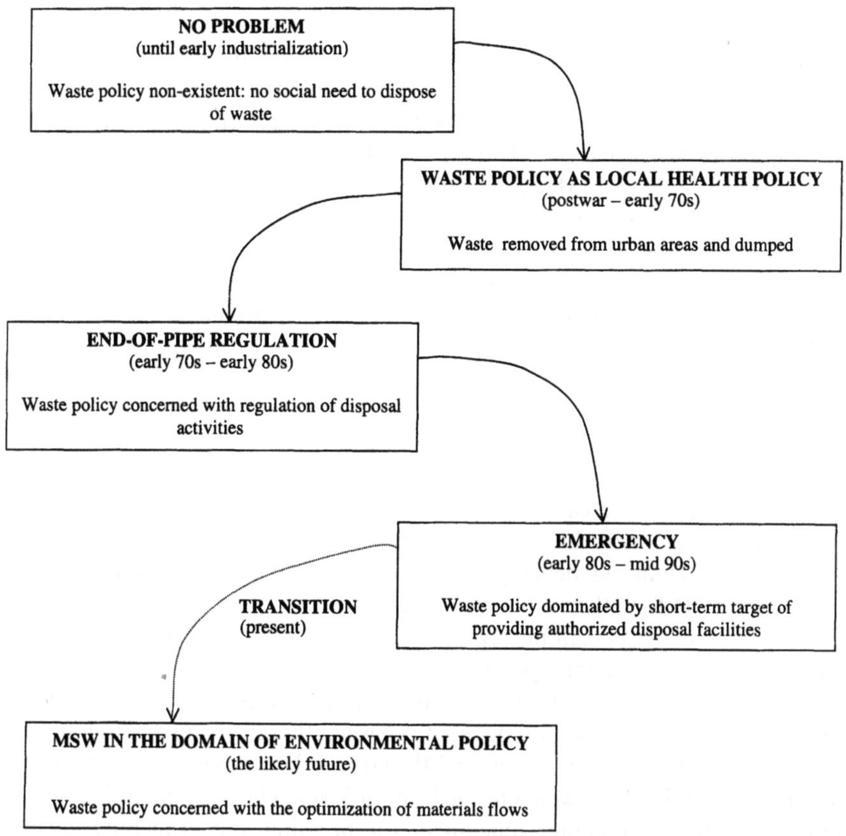

Figure 1. The evolution of MSW management regimes in Italy

Waste policies as local health policies

Although there are traces of regulations concerning waste management in laws and ordinances of the Roman republic and the medieval *Comuni*, it would be exaggerated to consider MSW as an area of systematic policy before the 1960s, and in many areas of the country before the 1970s. Before this date, waste was not considered a problem: quantities were low, and organic materials were acquired for soil improvement purposes by nearby farmers. There was still a substantial equilibrium between town and countryside, and the recycling of all other materials was almost total, as in the case of any poor and economically under-developed country. The urbanization process altered these equilibria and systematically began to pose health problems, enhanced by the fact that development had occurred very rapidly, without very much planning of the new urban infrastructures.

The setting up of a collection service, provided by the municipalities, occured quite early on: formally with the *T. U. delle leggi sanitarie* of 1934 and more specifically with the law 366/1941. This act made it the duty of municipalities to provide collection facilities for waste removal in urban centers. Earlier on, in 1931, another law had enabled municipalities willing to provide waste disposal services to finance them with a special tax.

After World War II the coverage of waste collection and disposal services was almost total in urban areas, even though it did not mean much more than the simple removal of waste from households. Many municipalities, especially in the North, decided to contract out the services to private firms. These were small, local enterprises whose advantage was almost exclusively lower labour costs and flexibility, particularly appreciated in a territorial model characterized by small centres and a dispersed population.

In larger urban areas direct management with its own employees and facilities prevailed. Increasingly, major towns adopted a slightly more sophisticated approach, with the creation of municipal enterprises, with their own budgets and accounting, developing the experimental model which had been in force since 1904 for other urban public utilities like gas, water and electricity. The service remained simple everywhere, essentially labour-intensive and concentrated on collection operations.

The issues concerned were basically the impact of waste on urban health and quality of life. The main preoccupation was, therefore, providing for the removal of waste from the houses and the streets. Technology was the simplest possible: home dustbins were emptied into a lorry. The only capital asset was represented by rudimentary machinery for pressing and crunching garbage before loading it onto the lorry. In this phase, the generation of waste was regarded as an indicator of wealth. New houses were built with waste chimneys in order to remove garbage as soon as possible. The quantity and quality of waste generated did not represent a particular problem, or was disregarded, along with the impact of industrialization and urbanization on the environment.

The market provided cheap and readily available disposal capacity almost everywhere. In the neighbouring rural areas, people were still quite willing to accept urban waste for soil improvement; the remaining waste was dumped in an indiscriminate way or incinerated in the open air. At the end of the 1960s, when quantities of waste began to rise and uncontrolled disposal began to produce visible negative effects, many cities were forced to find longer-term solutions. These barely modified the traditional model. In the majority of cases, holes in the countryside were enough. The larger towns acquired large areas for building landfills, which became available because the urbanization process involved the abandonment of rural areas, and massive quarrying works providing construction material for the urban areas left void space at their disposal[2]. Although facilities were sometimes owned and operated by the municipalities themselves, either directly or through contractors, in the majority of cases the private sector directly provided the disposal service, which was acquired by the municipality. In some cases, and especially in the early 1970s, rudimentary facilities were provided for incineration and composting (Table 1).

In this phase MSW policies not only could not be inserted into the frame of a 'national regime' of any kind, but also could hardly be considered an 'environmental' policy problem. Technical regulations, if any, were very few in number; quality standards were freely decided by municipalities, and the concept of quality itself was much more related to urban cleanliness and citizens' comfort than to the environment.

Table 1. Trends of the MSW management system (1970–1985)

Years	Trends in disposal industry	Trends in collection industry

1970–1985 baseline	– Disposal facilities were owned by private actors, selling their capacity in a market whose size is mostly local.
	  – Most of them derived from quarrying activities.
	  – Exception: facilities serving larger towns, often under municipal ownership.
	– Disposal price was set through open market transactions and was in general negligible with respect to collection costs.
	– In the 1970s many municipalities in northernn Italy built incineration facilities that had to be updated in the 1980s due to poor emission quality.
		– Collection costs represented 90% of total costs
		– Highly labour-intensive, not specialized. Mostly undifferentiated.
		– Emphasis was put on urban dimensions of quality (e.g. frequency, availability, quality of street cleaning).
		– Pressure for labour cost reduction. Solutions were mechanization, rationalization, abandonment of individual pickup technologies in favour of large street containers.
		– Poor integration. Virtually no firm was active in more than one Region.
		– Exception: producers of some capital goods such as treatment facilities
		– Segmentation of the market:
		  – Urban areas were almost always served by their own enterprises.
		  – Private collectors, especially in smaller centres, in general entailing lower quality for lower prices.
		  – Nearly half of the market was still served through direct municipal operation.

Emergency – MSW as a problem of public order

A new MSW regime focused on the supply of authorized disposal sites

At the end of the 1970s everything became more difficult. Increasing quantities of waste needed to be disposed of, while environmental concerns began to grow. Incineration plants were found to be responsible for producing dioxin, which came as a violent shock in a country still reeling from the experience of Seveso. Some facilities were quickly closed. The environmentalist movement came to the forefront. The new environmentalism had a 'leftish' character, with strong links with the protest movements of the late 1960s; these views were combined with a more traditional national environmentalism, much more concerned with issues such as landscape and cultural heritage than the environment as such. The 'Limits to Growth' debate contributed to present the waste problem to the general public essentially as a problem of wasteful use of raw materials. Utopian ideas of 100% recycling began to be shared by a vast audience, yet all this remained at the margin of policymaking.

Despite these internal pressures, the trigger for institutional development, as in most other issues of environmental concern in Italy from now on, arose as a result of European legislation. Dir. 75/442 was enacted by DPR 915/82, which started a new phase in MSW policy, anticipated by the legislation put through autonomously by some Regions (like Lombardia and Emilia-Romagna) in previous years.

The pillars of the 1982 legislation (Dpr 915/82)

- *authorization regime* for all relevant activities in MSW management (collection, operation of disposal facilities, transport, sorting etc). In order to obtain authorization, technical standards had to be met, financial guarantees had to be provided, and special features had to be verified.
- *classification of waste flows*: subjective, based on the origin and destination of of waste flows; the category of *residui*, identifying materials that could be used as input and could therefore escape waste regulations and duties of care
- *classification of landfill sites*: different categories of waste could be disposed of only in facilities of a certain class – while each class involved different technical and location standards. The relevant categories were 1 (for MSW), 2A (for inert materials), 2B (for industrial and commercial waste), 2C (for toxic waste).
- *technological and emission standards* for all treatment facilities and/or reuse of waste-derived by-products, such as compost
- *duties of care* to be respected in the use of waste-derived materials (e.g. compost)
- *regional waste disposal plans* aimed at devising a strategy for MSW management and promoting the development of disposal capacity
- *decentralization*: responsibility lay with municipalities, who had to comply with RWDP regulations concerning disposal technologies, and separate collection and recycling targets. In the case of objectives not being met, Regions and the Ministry of the Environment were given the power to provide the services directly (*commissariamento*)

Public powers were now concerned with the issue of regulating the environmental impact of disposal activities, in response to the substantial deregulation of the past. The objectives of waste regulation, environmental protection and the recovery of resources were added to public health and safety.

The dumping of waste was prohibited, unless in controlled sites licensed by the competent authority. This implied the setting of emission standards and technical regulations, and, moreover, the establishment of a licensing system for all waste-related activities (from handling to transport and disposal). Disposal sites were classified according to the categories of waste they were authorized to receive; for each category, construction and operation standards were set, together with precautions to be observed and areas that were to be excluded. The same applied to incineration and composting facilities.

A crucial distinction was made between MSW (where responsibility continued to lie with municipalities, though under a regional planning framework) and productive waste, for which producers were responsible. While the legislation did not offer any explicit priority ladder, it put emphasis on the minimization of the harmful effects of waste disposal, and mentioned waste reduction, recycling and recovery among the available options. Though not explicitly, therefore, the law suggested a strategy of abandoning landfills in favour of other technologies.

Another pillar of the 1982 legislation was the definition of waste, which is particularly troublesome in a country where recycling activities have strong traditions and the very concept of waste is difficult to define. Italy basically adopts a subjective definition, according to which waste is whatever a person decides to get rid of. New categories were defined: toxic waste was such if it contained one of a list of substances above certain concentrations; conversely, residues (that is, waste suitable for recovery) were defined in such a way that they could be considered equivalent to any other material[3]. The polluter pays principle (PPP) was intended to make the waste producer responsible for the total cost of waste disposal; with respect to MSW this meant that municipalities, as representatives of citizens, had the duty to be responsible for sound waste disposal and to bear the cost; and the right to adopt their own regulations and organizations for the purpose and raise a tax aimed at meeting costs[4].

Regions created in the 1970s in accordance with the Constitution of 1946 are the level of Government mostly involved in the administration of this and other areas of environmental policy. They are held responsible for all administrative aspects, and especially for licensing waste disposal facilities. The Regional Waste Disposal Plan (RWDP) was set up in order to achieve a coordinated system. The RWDP was aimed at ensuring a sufficient disposal capacity and adopting preferred technological strategies. RWDPs were indeterminate in their scope: there was considerable flexibility in setting priorities and technical standards, and allocating waste flows to treatment facilities. Nonetheless, municipalities remained responsible for all waste operations. They could fulfil this duty making use of a number of possible arrangements going from direct management with its own employees to the creation of special enterprises (*aziende municipalizzate*) or delegating to private operators. They could also voluntarily create associations with other municipalities (*consorzi*); in some cases Regions forced the creation of these associations, especially for the purpose of managing larger disposal facilities.

Implementation problems

With the exception of a few Regions in Northern Italy, Italian regions began to draw up, and issue RWDPs between the end of the 1980s and the beginning of the 1990s. Campania had the first approved RWDP in 1994: a draft issued in 1986 had never been approved by the regional assembly, and so never enforced. Sardinia obtained one in 1995.

Paradoxically, the effect of regulations on the supply of disposal services was completely opposite to that desired. Also due to the new regulations, people had become more aware of the harmful potential of waste disposal, and the siting of any facility for handling MSW was from now on fiercely opposed. At the same time, operators and technicians seemed to foster basically an end-of-pipe approach, looking for technical ways that could mitigate the environmental impact. Environmentalists criticized this type of strategy and put all the emphasis on recycling, which operators and official experts dismissed as ineffectual in terms of achieving highly demanding targets. Strict emission limits were indiscriminately placed on older plants and new plants: these were not strict enough to match more severe legislation in other countries, which failed to satisfy environmentalist opposition to new facilities, but strict enough to sweep existing facilities out of the market and entail very costly restructuring and restyling.

Much the same happened in the case of compost. Regulations for the use of this waste-derived fertilizer were both too complicated[5] and too loose with respect to quality standards; moreover, good composts (made from selected organic waste) were not differentiated from compost derived from mechanical sorting plants. As a result, most of this compost ended in landfills, or was given away for free, thereby ruining the market for better quality compost.

While the development of what were referred to at the time as 'complex technologies' was negative, the market supplying landfill sites began to dwindle and was no longer able to meet the demand, particularly in areas such as Lombardia. Rapid fortunes were made by those few who managed to recycle old quarries into waste disposal plants.

In terms of intention, the Regional plan was aimed at offering a convenient supply of disposal facilities by simply shifting responsibility from the municipal level to the Regional level, but this objective proved to be very difficult to achieve. Long-term studies were carried out, basically concerned with the analysis of waste flows, the design of supply areas and the planning of new facilities. The approval of the plans was repeatedly delayed. The need to reconcile too many different technical and political views led to a compromise: the compromise solution lay somewhere in the middle between that desired by MSW operators, namely incineration, and that favoured by environmentalists, namely recycling. An analysis of RWDPs issued during the 1980s shows a clear predominance of mechanical sorting plants aimed at end-of-pipe recycling (producing compost and refuse-derived fuels – Rdf). Very demanding targets for separate collection and recycling generally accompanied these measures, yet very few incentives were provided. Not surprisingly, longer-term solutions such as waste-to-energy plants were shelved, either because of the complexity of licensing procedures or because of the difficulty in putting together a sufficient number of municipalities so as to achieve economies of scale. The political cost embedded in any siting decision was high, and this made administrators very reluctant to back waste-related projects.

In our opinion, this poor commitment at the municipal level, together with the high sensitivity of air pollution issues and the spectre of dioxin, are the main causes lying behind the limited recourse to incineration. Municipal officials, and especially politicians, had no incentive to raise such an unpopular issue, given that some other solution, at least in the short term, though at a rising cost, was available.

The fact that operators, producers of technology and 'official' experts were quite unanimous in support of 'safe incineration' was not enough to dampen local resistance, but it was enough to provide operators with an alibi for their incapacity to deal with the waste avalanche. At the same time, MSW operators did not welcome the strategy fostered by regional plans, perceived as wishful thinking with regards separate collection targets, and technically unfeasible with regards end-of-pipe recycling, and, as a result, did not cooperate very much in the setting-up of the technology. Those few plants that were set up at the time almost always had technical weaknesses. The expected development of the market for compost and Rdf did not occur automatically: it required a coordinated effort whose complexity had been dramatically underestimated.

Though dominated by a rigid command-and-control approach, RWDPs, in the end, turned out to be superficial and the product of wishful thinking. The available policy instruments were strong enough to ban undesired activities but too weak to foster a positive action on the implementation side. Objectives were at best individuated, but no coherent action was articulated as a driving force behind the development of the system. The need to develop the industrial organization of the waste *filiére* was dramatically underestimated. Enforcement measures were harsh, but often inapplicable. The control system was inefficient, dispersed and unevenly distributed among the Regions, leaving space, as a result, for illegal activities.

In the end, most of the initiatives generated by the RWDPs simply remained on paper. The waste crisis exploded in several areas in Northern Italy, bringing in its wake social conflict. Control of the emergency is achieved painfully through acrobatic and inconclusive negotiations, long-distance waste transfers, and by opening new landfills by-passing ordinary decision procedures. Only in a number of smaller towns with a strong tradition of good local administration did some incinerators and well managed landfills guarantee a safe disposal at reasonable prices.

Table 2. Trends of the MSW management system (1985–1995)

Years	Trends in disposal industry	Trends in collection industry
1985-1995 FROM REGULATION TO EMERGENCY	● *Religious war* for technological choices: incineration, sorting and recycling were seen as alternatives and were sustained by precise industrial and environmental lobbies. ● Ownership of disposal facilities - and especially the new ones - was increasingly held by the public sector, which later could contract out plant operation to private firms. ● Incinerators and mechanical sorting plants were always owned and nealry always operated by municipal enterprises, municipalities and syndicates. ⇒ RWDPs often forced or promoted the creation of municipal syndicates for the operation of new plants ⇒ Exception: some development of build-operate-transfer techniques, with the private operator never bearing the risk of the investment; operation was required basically for training municipal staff and/or for ensuring a better technical commitment in the project and construction phase. ⇒ New trend: l. 9/91 provided a strong incentive to alternative power production technologies, particularly if "renewable". This attracted new entrants in the WTE *filiéres*: other urban utilities, private operators able to use waste-derived materials in own combustion processes, investors in WTE projects developed by municipalities. ● Nonetheless the private sector still remained important: ⇒ outside congested areas, the "normal" way of supplying disposal capacity was through private plants, ususally abandoned quarries ⇒ in some congested areas private operators were encouraged to create new plants under short-term planning arrangements. Emergency procedures often allowed them to by-pass ordinary authorization processes ⇒ private operators still remained dominant in the market of industrial and commercial waste; this market was often the final destination of MSW flows, especially when emergencies forced people to look for solutions outside the RWDP. ● Disposal strictly separated from collection. Different firms dominated the 2 sectors ⇒ A large number of plants were operated and sometimes owned by small local firms with limited technical capabilities ⇒ Unclear boundary between categories of waste encouraged illegal practice. ⇒ Exception: large private operators in the collection industry also had own capacity, but were not normally free to use it for supplying the areas where they collected waste due to RWDP restrictions ⇒ Exception: a number of municipalities with own treatment plants managed services directly or through own enterprises Plant operation and selling of disposal capacity could be separate activities. Financing mostly came from public funding both at the national and the regional level. Exception: private plants under short-run planning arrangements, generally enjoyed concession contracts with monopoly rights and regulated prices (landfills only).	● While Regional plans lagged behind, tension was created in collection services: many new services needed to be set up (e.g. separate collection), needed to look for expensive and precarious disposal solutions. Day-by-day attitude. ● Increasing difficulties due to traffic, etc. ● Ineffectiveness of planning policies stimulated many operators to devise own solutions (e.g. treatment plants) searching for integration with neighbouring markets. This strategy was successful only in a limited number of cases, generally public enterprises ● Private operators tried to increase their market position but only with partial success: ⇒ aggressive penetration of multinational firms (WMI, BFI...), mostly through the acquisition of existing operators with their market shares ⇒ competitive strategy aimed at forcing public establishments out of the market, but only a very small number of municipalities - often very small - decided to "privatize" the whole service ⇒ In the part of the market traditionally covered by private operators, quality requirements remained very low and the main problem was avoiding extra-costs. Larger firms were often defeated by local subjects, some authors have argued that an adverse selection problem occurred. ● At the cultural and political level, the debate about privatization employed an aut-aut rhetoric: public and private operation were perceived as antagonistic models, with supporters of both claiming superiority. ● Emphasis on corporate integration within the same organization. Operators still perceived labour-intensive activities, and especially undifferentiated collection, as their core-business.

In the meantime in Southern Italy and in most of Central Italy, with the exception of Tuscany and Marche, almost nothing happened; RWDPs were sometimes drawn up, seldom approved, and never enforced; waste continued to be disposed of in hundreds of uncontrolled landfills, mostly privately owned, at very low prices. The pressure environmentalists could bring to bear in those regions was negligible.

The huge landfill capacity (there was virtually no control on the quantity, and quality, of the waste swallowed up by privately owned landfills) at very low prices in the South started attracting a waste flow from the North. In some areas, mostly in Lombardia, but also in Emilia, Veneto and Tuscany, landfill prices went up to 100–150 Euros/tonne, while in Southern Italy the cost was a few Euros/tonne and the price 20 Euros/tonne. The transport of waste to Southern Italy became convenient, and the owners of the landfills in Southern Italy, often connected to the Mafia, made huge profits. This waste flow from the overcontrolled areas in Northern Italy, where the incapacity of the planners and the pressure of the environmentalists prevented building new facilities, to the Mafia-controlled landfills in Southern Italy marked the failure of waste planning in Italy.

Thus, the role RWDPs were able to play in concrete terms was just a precarious regulation of internal waste flows, and whose main achievement was to guarantee solutions to emergencies and mediate between areas possessing disposal capacity and areas needing it. Concrete planning policies during the 1980s and early 1990s had much shorter-term purposes than in the original aims: control of rents through the regulation of disposal prices; emergency ordinances in order to allow centers without disposal capacity to use other plants; regulation of waste traffic; promotion of new initiatives through financial grants and the supply of consulting services through public agencies.

Despite RWDPs, landfills continued to dominate the scene. In Northern Italy, the majority of waste is landfilled in controlled and authorized sites, while in the South waste is mostly dumped in small landfills, virtually outside any control. During the 1980s, due to the increased control, and to the enforcement of waste planning, small municipal landfills disappeared in Northern Italy, while in the South most of the 3000 *comuni* still ran their own landfills. The environmental quality of facilities was generally not very high, and definitely bad in the case of older plants. Not more than 10% of MSW was incinerated or treated in mechanical sorting plants; the output of these facilities very often ended up in landfills because of the lack of market opportunities.

The most evident, and perhaps durable, effect of the regime, however, was the vertical segmentation of the market: collection and disposal become two almost completely separated activities. Collection meant collecting waste following municipal regulations, and under municipal authorization. Disposal meant running one of the facilities authorized by the Region, receiving waste flows scheduled by the RWDP, at a price set or negotiated at the Regional level, and with the obligation to accept only waste produced in certain areas. Though there were some areas of uncertainty (like waste compression and transport facilities, and increasing all activities concerned with waste valorization), this distinction was quite straightforward. Even if the same firms operated in both markets, they operated under an agreement with different authorities and did not necessarily serve the same municipalities with all services.

The picture of waste management in most of Italy was quite hazy: in the critical areas of Northern Italy the emergency was continuous, as seen above; nonetheless, in many cases the waste stream coming from these areas made the equilibrium in neighbouring areas precarious, where a more or less satisfactory waste removal together with a controlled disposal into available landfills had been achieved and maintained. Older landfills, which had permitted them to escape crises up until then, were rapidly being exhausted or were finally forced to close[6]. Emergencies in other areas also created pressure. Moreover, in some areas of Southern Italy, the flow of waste coming from the North started to pose problems: in Puglia and Campania ordinances preventing operators to receive waste from other areas were issued at the beginning of the 1990s. Thus, even Regions whose waste policies were still lagging behind were forced to dedicate greater attention to MSW management. Some of them began to introduce limitations to waste imports. The situation was quite difficult to control; new actors increasingly dominated the scene from the backstage: intermediaries, firms specializing in long-distance transport, and brokers.

Waste scandals call for further regulation and increase the diffusion of separate collection

Some blatant cases of illegal disposal were discovered and created serious public scandals. The most famous one regarded the so-called 'poison ships' transporting toxic waste to Third World Countries: when this traffic was discovered, many of these ships were sent back to Italy and an emergency plan had to be adopted in order to receive them back (no harbour was willing to let them dock) and dispose of the waste. To cope with this situation, a number of measures were adopted at the end of the 1980s: all characterized by very heavy and complicated administrative and authoritative instruments, whose success was only partial (Table 3).

Firms wishing to operate in the waste industry were required to be inscribed in special registers, authorizing them to operate in a particular Province. The Provincial dimension of the authorization was perhaps aimed at reducing 'waste tourism', but the solution proved not only to be ineffective, but also placed a number of burdens on most reputable operators, which finished up by favouring illegal operators. Complicated bureaucratic requirements were also introduced, with the aim of monitoring the movement of waste across company boundaries and operators. However, the final outcome was not necessarily optimal: firms wishing to respect the rules were oppressed by an enormous number of regulations, often in contradiction with each other, and a strong incentive was offered to illegal activities.

The weakness of pollution control made prosecution of offences very unlikely, and the severity of punishments, including prison, was not a sufficient

Table 3. New waste regulations introduced during the 1980s

- *Regulation of waste transportation activities*, brought into an authorization regime; the fact that authorizations are mostly regional or even provincial enhanced the local character of the market, on the one hand coherently with the proximity principle, but on the other creating artificial barriers to competition.
- *Introduction of many special and ad-hoc norms* for particular waste fractions, especially toxic and hazardous waste
- *Obligation to devise separate collection schemes* and later to meet recycling targets
- *Institution of a landfill levy*, which is currently around 13 E/tonne
- *Obligatory Syndicates* aimed at promoting recycling of packaging waste (only containers for liquids), used oil and batteries (see section 4)
- *Constraints to decentralization*: while municipalities remained responsible for MSW management, RWDPs increasingly tried to dictate priorities, technical and operational standards and also technical choices. They also tried to promote or even to force the creation of municipal associations and syndicates for the construction and operation of new plants and the coordination of separate collection and recycling

discouragement. Brusco *et al.* (1995) talk about 'capture of control' to depict a situation in which a clear asymmetry is present between the level of regulatory attention devoted to the largest and most well-known operators and the mass of 'cowboys' operating at the boundaries of legality. The market therefore had evident features of adverse selection. The best initiatives were discouraged, since the bad reputation of anything having to do with waste was widespread; in this void of official initiatives, a silent, less visible regime was left in operation. If not illegal, this regime could at least be defined as irregular, dominated by emergencies, very short term and precarious; and, in any case, far from the achievement of waste policy targets.

Though most of these measures regarded industrial and toxic waste, the MSW industry was also part of the picture. First of all, the boundaries were not always clear. As soon as ordinary planning procedures blocked the regular disposal supply, many cities were forced to look for landfills elsewhere and therefore enter the spot market dominated by the same intermediaries and brokers, often sending MSW thousands of kilometres away without much control. Second, the general distrust surrounding waste disposal activities also had an effect on the social acceptability of management practices pertaining to the MSW cycle. Third, illegal activities involved operators in the MSW industry as well as in the other fields of waste disposal[7]. Fourth, and perhaps most importantly, the wide public attention given to waste policy was reflected in the increasing influence of environmentalists at the legislative as well as the planning level. The most evident effect of this was the increasing emphasis put on recycling. National legislation forced municipalities to provide separate collection for a wide variety of MSW fractions.

Regional plans required even more demanding targets for recycling. Again, however, the incentives provided to municipalities were too weak, and enforcement measures were absent. Thus municipalities reacted quite slowly, and, in many cases, they simply limited themselves to a formal fulfilment of this duty (e.g. by putting one glass or paper bin near the schools and nothing more). Municipal operators were not willing to be burdened with extra costs, which could not be covered due to the political unfeasibility of increasing the waste tax. Nonetheless, in some cases incentives were higher. This happened where landifll prices had reached and maintained very high levels for long periods. There is some evidence that the increase in separate collection rates and therefore, average costs, during the 1980s are correlated with the rise in landfill prices (Massarutto, 1998). A second typical situation occurred when municipal operators provided separate collection essentially in order to develop a 'green image', which could later be exploited in order to gain acceptance for other projects.

Private operators, for their part, were quite willing to develop separate collection schemes provided that this meant an enlargement of the market and new business opportunities; the most far-sighted of them increasingly saw the development of separate collection and the organization of a territorial network of facilities for sorting and valorizing these waste flows as a way of strengthening the market linkage with local authorities.

A third situation occurred when pressure from citizens and environmental associations persuaded municipal officers of the social desirability of some 'greening' of MSW management. Separate collection was valued essentially for its educational and socializing capabilities, and very often was managed in cooperation with local NGOs, social cooperatives, and so on.

Across the 1980s and the 1990s, separate collection made some progress. Aggregate results were negligible, not more than 2–3% on average, with better performances (around 10–15%, in a handful of cases), nearly always in the North and in medium-small towns[8]. Yet there was alrady an *avangarde* of municipalities, generally small and located in the North, aiming at far more ambitious targets.

The 1988 decree also provided for the institution of Obligatory Syndicates of producers and users of packaging and batteries, following the earlier and in many respects successful experience in the field of mineral oils. This form of organization anticipated some of the features that the packaging sector would acquire later in many European countries following the DSD (see below).

Across two regimes: triggering elements and issues

Despite this evolution on the regulatory level, the emergency phase reached its apex at the beginning of the 1990s. The waste crisis was gradually but inescapably extending to the whole country. Building any new facility was practically impossible, if not at astronomical prices. Although a minimum target had been reached, with MSW disposed of in authorized plants and in a controlled way, landfills remained largely predominant and waste transfers between Regions were significantly numerous. Though most Regions had issued their RWDPs, the structure of the management system seemed incapable of implementing them, and in most cases was definitely unwilling to do so. Financial resources were available for covering at least the initial capital cost, yet only in a handful of cases, generally through the action of a public company, was it possible to bind a sufficient number of municipalities to the project in order to guarantee the necessary cash-flow. Even though grants from the general budget were available, the capacity to canalize this budget into technically effective projects was extremely poor, and often hampered by corruption scandals and capture of planning decisions from the side of equipment suppliers. At the same time, while the achievement of the targets set in 1982 was still largely incomplete, the EU initiated a new policy based on radically new principles such as self-sufficiency and the priority ladder.

Social awareness of the waste problem, thanks to the initiative of environmental associations but more generally because of its great visibility, raised MSW to the first place on the municipal agenda. Environmental campaigns and volunteerism had an important educational effect on public opinion; at the same time, the emerging EU framework and the example set by other countries was used by environmentalists in order to gain a wider audience and acceptance in the decision-making process. Operational costs rose, putting strong pressure on municipal officials who wanted to avoid sharp increases in the waste tax. Together with the larger operational scale of the 'new' waste-related activities, this increased the pressure to delegate the activities. This new tendency found a MSW industry that was itself undergoing a process of development: concentration and technological evolution in the private sector (especially after the aggressive penetration of foreign multinationals like WMI and BFI) was matched by a more entrepreneurial strategy on the part of municipal enterprises, which tried to geographically extend their market area from urban centers to nearby metropolitan areas or the countryside.

The transition was very slow. Supporters of the status quo and of the traditional end-of-pipe approach were still strong, while municipal control of the bulk of the waste industry often represented an obstacle to change, since municipalities were generally reluctant to give up control over the waste issue. While evolution of the management framework required an increase in the operational scale, it was normally difficult to find sufficient consensus regarding a common waste management strategy at the provincial level. Major towns were perceived from outside as would-be waste exporters, and nearby municipalities were unwilling to be drawn into their orbit, unless they themselves had precipitated into the waste crisis. Incentives provided by RWDPs for the creation of syndicates, including the power to force their creation, were only rarely successful.

The painful way out of the emergency: a new regime for MSW management

New legislation

Around the mid 1990s, the situation seemed mature for a change in the national regime. As in many other cases, the European input was decisive in forcing governmental action and displacing internal opposition. The new framework was introduced by a ministerial decree, taking advantage of an opportunity offered by the community law of 1996, which facilitated the legislative process enabling the government to directly implement European directives whose implementation was lagging behind. The new Minister of the Environment, from the Green Party, exploited this opportunity with the Dlgs 22/97, which had been drafted and widely discussed and consulted during the previous year. This new piece of legislation completely substituted the previous system based on Dpr 915/82. It reorganized the whole regulatory and management system, introduced a radical reform of financing patterns, and, finally, significantly widened the range of policy instruments available to waste policy-makers, with a decisive shift in favour of economic and voluntary instruments.

Table 4 summarizes the main innovations introduced on the regulatory side implementing the four principles of European policy for MSW. With respect to the self-sufficiency and proximity principles, the chosen approach was quite flexible. Relevant territorial units were 'optimal management areas' (OMA), which roughly correspond to the Provinces (a 100 in the whole country, for an average population of 500 000), but the export ban was tempered, since it regarded only raw MSW. Treated waste, e.g. waste arising from sorting plants, could therefore escape this rigid limitation. Regional government was given some freedom to enlarge OMAs or to regulate waste transfers between them. Regions were also free to negotiate interregional waste transfers, requiring the consensus of both parties involved.

The ladder principle was interpreted with emphasis on flexibility and complementarity of technologies. A ban on landfilling untreated waste, which could be seen as a moderate version of the French concept of ultimate waste, was to be introduced in the year 2002. Separate collection and recycling targets were determined for each OMA and for each material, as shown in Table 5. Between these extremes, there was freedom to choose any desired technological mix.

On the management side, the law required all municipalities lying within each OMA to set up a coordinated or even integrated management system. Again, Regions had considerable freedom in defining the contents of this regulation. Therefore, municipalities, while maintaining formal responsibility, were in practice forced to accept a collective establishment with responsibility for the whole vertically integrated MSW management system. This provision was intended to mitigate the negative effects of the previous dualism between collection and disposal.

The promotion of separate collection, recycling and waste avoidance was effectuated, on the one hand, by the reinforcement of monetary disincentives

Table 4. The Italian approach to MSW management principles set up by EU 91/156 Directive (Dlgs 22/97)

Principle	Implementation in Italy
Proximity	– MSW cannot be disposed of outside Regions – Does not apply to industrial waste (including processed MSW)
Self-sufficiency	– Optimal Management Areas (OMA) should be determined; within each OMA, municipalities should move towards a coordinated management system, if not total integration – Each OMA should be generally self-sufficient – Flexibility allowed (e.g. no total bans on MSW exporting between OMAs but strong disincentives through environmental taxes and extra charges; possibility for OMAs to merge) – More rigid for 'final waste' than for intermediate activities
Polluter-pays	– Landfill tax (10–15 Euros/tonne) – Full cost pricing for waste disposal services – Differentiation of charges according to waste production and environmental policy objectives – Responsibility for the cost of waste valorisation gradually shifted onto producers (e.g. packaging)
Priority ladder	– Recycling targets (see Table 2) – Prohibition to landfill waste that has not been properly valorised or treated – Wide adoption of producer responsibility (e.g. packaging, oils, cars, household electrical appliances – Strong emphasis on waste prevention: – Capacity to prevent waste as a discriminating criteria for MSW management contracts – Precautionary principle

Table 5. Recycling targets imposed by the Dlgs 22/97

Separate collection (each territorial unit)	At least 15, 25 and 35% within 2, 4 and 6 years
Packaging waste (recycling or energy recovery)	50% min, 65% max
Packaging waste (average recycling)	25% min, 45% max
Packaging waste (recycling rate for each material)	at least 15%

(e.g. the landfill tax) and, on the other, by the obligation to meet recycling targets; there was also a clear shift towards a mechanism of producer responsibility. This was foreseen not only in the case of packaging but more generally for every identifiable waste flow, through voluntary and negotiated agreements with the respective industrial *filières*. The next candidates were electrical appliances, household appliances, used tyres, used furniture, and demolition waste among others.

Lastly, a decisive and more rigorous application of the polluter pays principle was introduced: the old waste tax was substituted by a waste tariff paid directly to the operator and directly correlated with waste generation and adjustable in order to foster participation in separate collection schemes.

Planning styles and management systems

A parallel development, partly a result of the new legislative framework for MSW, partly independent, regarded the organization of the management system at the regional and municipal level.

Regions have begun to abandon the authoritative planning style dominating the 1980s; while aiming at control over disposal capacity they have tended to abandon their proactive role and concentrate on the regulatory side. Operational decisions, such as disposal technologies and siting of facilities, are increasingly left to local initiatives: the planning problem has become the search for an appropriate incentive system, which is increasingly found in a combination of direct responsibleness, economic instruments and voluntary/cooperative approaches. On the operational side, it seems there has been only a very slow evolution. The traditional segmentation of the market between different management systems (direct management, municipal enterprise, delegation) has shown very little differences over time. The situation of 1994, described in Table 6, was only marginal different in comparison with 10 years before.

Table 6. Operation of MSW services in Italy in 1994 (% of population)

	North West	North East	Center	South	Italy
Direct management	12.4	30.4	48.9	62.2	39.2
Municipal firm	30.1	28	36.8	12.9	25.9
Contracted out	36.9	22.3	9.4	16.7	21.8
Syndicate	9.2	7.9	0.4	0.1	4.2
Other	11.5	11.4	4.5	8.2	8.9
Total	100	100	100	100	100

Source: Iefe-Eco and Eco 1997b

A closer look, however, shows a more subtle and perhaps subterranean evolution. While municipalities continued to hold responsibility, they increasingly cooperated with each other, at least for the carrying out of specific phases; while new patterns of labour division characterized the industry, with higher specialization and horizontal cooperation. Thus in Milan the form of management remained the municipal enterprise; yet all the new phases of the service were sub-contracted out, and for the new sorting and treatment plant, a joint venture between Amsa and five private partners was created. Many cities developed a strategy of horizontal integration: municipal enterprises operating in different public utility sectors were merged together and often privatized; their operating branches were often separated, implementing an outsourcing policy, and selling specialized services to other towns. Private firms were evolving as well: contracting out was no longer perceived as a crude alternative to public management; instead, the multiple opportunities for combining the two, at different levels, were being increasingly explored. In short, the last 5 years has witnessed a rapid process of development of the MSW industrial *filiére*, together with a slow but unavoidable transfer of responsibility and decision powers from planning officers to operators (Table 7)

Table 7. Trends in the MSW management industry (late 1990s) (towards the 'green target')

- *Complimentarity of technologies* is recognized by all actors. No more religious wars, increasing agreement about the spectrum of available alternatives. Differences among 'technological parties' regard the different emphasis to be placed on different elements of the system rather than placing them in contraposition.
- Increased vertical integration between management activities at the municipal level (collection, sorting) and disposal/treatment, leading to an increasing role of the MSW operator in strategic decisions concerning disposal and management options
- Increased territorial integration has led to a progressive abandonment of direct municipal management
- Increased monopolistic aspect of the market; control of disposal capacity increasingly represents a decisive asset
- Facilities authorized for industrial waste (able to receive what results from MSW after sorting and separate collection)
- Sorting plants, pre-recycling plants, etc
- Pressure to reduce costs, increasing technical complexity and growing market dimension leads to new opportunities for private operators:
- Total or partial privatization of municipal utilities (often: mixed companies, with majority of shares held by municipalities)
    - Joint ventures between municipal utilities and private partners possessing technical capabilities, available disposal sites and/or financial resources
    - New investment (project financing, mixed venture companies)
    - Outsourcing and decentralization of operational activities (especially labour intensive)
    - Increasing need for operators specialized in specific activities
- Market development at the boundaries of the waste sector:
    - New actors on the demand side (packaging consortia, other industries developing *filiére* approaches such as automobiles, used tyres, household appliances)
    - Valorisation of specific waste flows (e.g. packaging, green composting)
- Shift of relevant skills:
    - Not only the capacity to provide low-cost solutions to labour-intensive problems but
    - Capacity to organize a complex network of activities, increasingly carried out by external firms and specialized operators

The Italian MSW regime at the end of the 1990s: MSW in the domain of environmental policies

Waste collection and disposal in Italy: basic data, main issues and trends

In broad terms, the quality of data concerning MSW collection and disposal in Italy is rather poor and unreliable. Only some very general issues can be addressed. The tradition of municipal responsibility for collection and disposal facilities creates dramatic difficulties for coordination efforts. Moreover, the country is very difficult to categorize both with regards to patterns of waste generation and to patterns of collection and disposal. Problems are very different in urbanized and rural areas, North and South, small, medium and large centres.

The quality of databases has slightly improved in recent years, both because of increased social and political attention being paid to waste-related issues and because of a wider involvement of Regional and National authorities in the field. However information remains dispersed, fragmentary and non-homogeneous. Special care should be paid to criteria used for classifications, which still diverge according to the different sources and which are not necessarily compatible with European standards. Only the recent Dlgs 22/97, enacting EU directives 91/156, 91/689 and 94/62, adopts a classification system which is coherent with these standards. With these *caveats* in mind, we can try to summarize the situation as follows, using information available from the most recent institutional databases and information provided by the waste collection and disposal industry.

In 1994, the production of MSW in Italy was estimated at 26.9 million tonnes (Table 8), corresponding to an average production of 460 kg per inhabitant. From 1991 to 1994, waste generation grew by 13.5%, 2.4% more than GNP; between 1994 and 1997, figures seem to be stabilized, varying between 26 and 26.6 million tonnes (Osservatorio Nazionale Rifiuti, 1999). Table 9 provides a breakdown of the average composition of MSW flows in different areas, showing significant differences between the North and the South with respect to aspects like organic content, calorific power, and packaging waste. Given past trends and likely evolution in the near future, a medium-term average breakdown is presented in Table 10, showing a rapid decrease in organic materials and a fast increase in caloric power. This trend, as in most industrialized countries, is mainly due to the growth of packaging waste, plastics and paper, and is strongly related to consumption patterns.

Table 11 shows the progress made by separate collection during the 1990s. Waste recovery from separate collection triplicate between 1990 and 1997 (from 0.7 to 2.1 million tonnes, corresponding to nearly 8% of total MSW). Paper and glass represent more than 65% of the total. The figure of 1997 also contains 0.6 million tonnes of organic waste used for green compost production.

Landfill remains the most diffuse option for disposal. Approximately 80% of MSW is landfilled, using facilities whose average quality is improving, but still doubtful in many cases, especially in the South (Figure 1).

Table 8. Generation of waste in Italy in 1991 and 1994 (thousand tonnes)

Waste	1991	1994	Variation %
Municipal	20.033	22.679	+13
Assimilable	3.166	4.232	+34
Special	19.127	19.464	+2
Inert	12.337	14.311	+16
Hazardous	3.246	2.709	–17

Source: Legambiente (1996); Ministry of the Environment (1997)

Table 9. Average composition of MSW flows in Italy in 1992 (%)

	NorthWest	NorthEast	Center	South	Islands	Italy
Chemical-physical data						
Water	40	41	43	49	50	44
Combustibles	36	35	33	29	30	33
Non combustibles	24	24	24	22	20	23
Energetic power (Kcal/kg)	1510	1410	1330	1020	1128	1300
Composition						
Paper and wood	25	23	23	16	20	22
Plastics	9	7	6	7	8	7
Metals	3	3	3	2	2	3
Inert	8	9	9	5	8	7
Organic	37	42	41	49	49	43
Indistinct	18	16	18	21	13	18

Source: Ministry of the Environment (1992)

Table 10. Average composition of MSW in Italy (medium term forecast)

Glass	Paper	Plastics	Metals	Organic	Other	Caloric power of raw MSW (Kcal/kg)
10	30	10	5	30	15	2400

Source: Kaulard and Massarutto (1997)

Half of the remaining amount is treated in mechanical sorting plants while half is incinerated. There are significant differences between Regions. Thus, while Regions such as Emilia-Romagna, Lombardia, Umbria, Toscana, Friuli-Venezia Giulia and Trentino-Alto Adige show between 60 and 70% of land-filled MSW, in other Regions this figure is around 100% and even more if we take waste imports into account.

In 1997 there were 38 incineration facilities, with a capacity of 7.7 million tonnes/day. These plants treated nearly 1.8 million tonnes, recovering a total of 430 000 MWh of electric and thermal energy. Incinerated MSW was 6.6% of the total. More than half of incineration capacity is concentrated in two Regions (Emilia-Romagna and Lombardia)

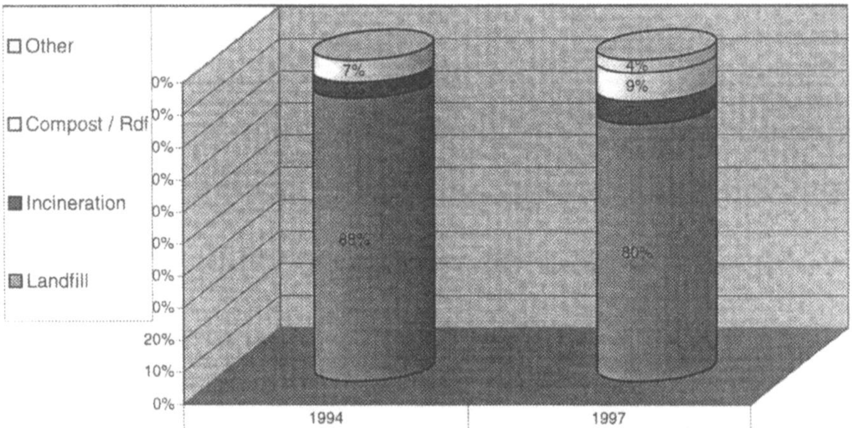

	1994	1997
Other		4%
Compost / Rdf	7%	9%
Incineration	5%	7%
Landfill	88%	80%

Figure 2. Final destination of MSW in 1994 and 1997.
Source: Ministero dell'ambiente (1998): Osservatorio Nazionale Rifiuti (1999).

Another 9.4% was sent to mechanical separation plants involved in the production of compost and RDF, for a total of 30 plants. However, as we have seen before, the amount of waste recovered in the case of this option is highly questionable. At present, nearly all RDF produced has never been used for energy production and is stocked 'provisionally' near the plants. The difference between these temporary stocks and open-air landfills is difficult to recognize. The use of RDF in industrial processes (like cement production and thermo-power generation) has long been advocated and proposed. A number of new initiatives has been announced, and even ENEL – the national electric utility – has shown interest; yet so far none of these projects has gone beyond the experimental phase. Compost is sometimes sold or given away for free for agricultural uses, yet in most cases no use has been found except for the recovery of landfill sites. Despite the improvements in sorting processes and management which has occurred over 10 years of learning-by-doing, this technology should be considered much more as a way of reducing the impact of landfilling operations than as a recovery option. At present, the dominant technological approach is to consider this technology as an intermediate phase, aimed at creating more specialized valorization opportunities at later stages: thus the production of Rdf is mostly aimed at obtaining a dry waste fraction with higher and more constant caloric power, while a destination for compost is seen in the reclamation of contaminated land and other similar activities.

These figures hardly reflect a situation of equilibrium. With respect to treatment facilities, projects currently under development (at any stage) are likely to drive non-landfill end-of-pipe technologies up to 25% of the total (Ausitra-Assoambiente, 1996). Medium-term forecasts show capacity for 3.3 million tonnes in incinerating plants (21 facilities under construction), 3 million tonnes in mechanical sorting plants (26 facilities under construction) and another 1 million tonne in 'green compost' facilities (15 facilities under construction). Energy recovery could reach 1 million Mwh from incineration only, plus the amount recoverable from Rdf.

Recycling performance has reached some 8% of the total amount of MSW. Separate collection schemes are now widely diffuse, especially in the North. Figures provided for different materials are not very reliable, since methods of calculation are not always homogeneous. Critical aspects regard the mixing of MSW and industrial waste flows, which completely alters the results for materials such as glass and plastics.

The Italian industry has a long-lasting tradition in the use of secondary raw materials in new productive cycles, as witnessed by the consistent flow of imported waste glass, paper, plastics etc. (Table 11). In fact, leaving aside the critical phase of the early 1990s, recycling capacity is at present sufficient to guarantee a stable market for separately collected materials. However, it is doubtful whether this assessment will continue to be valid in the future, if compulsory recycling targets set by the EU regulations implemented by the Dlgs 22/97 (see Table 5) are to be met. A further doubt regards the quality of materials recovered from MSW flows; in fact the tradition of reuse and recovery we are talking about has regarded mostly residues of industrial processes (much more

Table 11. Development of separate collection and recycling (1991–1997)

	1991	1997
Plastics		
Municipalities served	361	3334
Population served (million)	5	33.8
Quantity collected (thousand tonne)	2	>34
Glass		
Municipalities served	5000	>5100
Population served (million)	44	>45.5
Quantity collected (thousand tonne)	543	700
Industrial glass + other (thousand tonne)	223	300
Imports (thousand tonne)	18	80
Aluminium		
Municipalities served	1437	n.a.
Population served (million)	13	n.a.
Quantity collected (thousand tonne)	4.7	7
Paper		
Internal collection (thousand tonne)	1902	2339
Imported wastepaper (thousand tonne)	689	1000

Source: Legambiente (1997); Osservatorio Nazionale Rifiuti (1999)

homogeneous in qualitative terms than MSW). However, this tradition could turn out to be very useful for the implementation of the packaging directive, particularly regarding secondary and tertiary packaging.

Traditional models of collection use street containers and/or sacks. Particularly during the 1970s, a core strategy on the part of the authorities responsible for collection (and especially municipal firms) was geared to the mechanization of collection, with the adoption, wherever possible, of capital-intensive technologies (e.g. rounds with only one operator). However, this strategy has been curbed by territorial problems (e.g. narrow streets in historical centers, conflict with traffic and parking) and employed mostly in less densely populated areas.

Separate collection traditionally employs street containers for single materials, but more innovative schemes are growing (multimaterial, door-to-door, separation of organic and recyclables), as well as services 'tailored' for specific purposes (*dechétteries*, special services for toxic or harmful substances, special schemes for commercial and light industrial waste, etc). These schemes have ended the experimental phase, and are are at the root of the outstanding recycling performances we have reported.

A growing number of areas are also interested in policies aimed at intercepting 'hybrid' waste flows: that is, waste that used to be included in the public service, but is differentiated from household waste on marketable grounds: e.g. commercial waste, the output from public park maintenance, garden waste, catering waste, fruit and vegetable markets, etc. The trend clearly seems to be an increase in the complexity of the collection system and perhaps a climbdown in capital intensity. The number of local initiatives is multiplying and significant results have already been achieved. These initiatives may not have made much impact on national figures, but they are not to be underestimated in terms of importance.

The average figures provided hide the great lack of homogeneity in the country. Recycling rates approaching 30% are to be found in a large number of small, medium and even large centres. Even higher recycling rates (in some cases reaching 70–80%) are achieved in a number of centres, particularly districts of small municipalities, which are very diffuse in the North-East of the country. The environmental association Legambiente each year gives a public prize (*Comune Riciclone*) to the municipality achieving the best performances. Figures released by this association show that very ambitious recycling policies are already in place in quite a significant number of municipalities.

The following box outlines some key features characterizing different Italian centres with respect to MSW strategy. The most important thing that emerges is the great variability of situations, on the one hand, and the very rapid pace of transformation, on the other. Waste handling options, management organizations and key actors vary according to a mix of variables: structural data (size, climate, proximity to areas with problems), availability of own treatment capacity, planning style (authoritative vs. cooperative, tradition of autonomy, propensity to horizontal cooperation).

Some typical Italian MSW management schemes

Milan (capital city of the Lombardia region, with 1 500 000 inhabitants; 3 000 000 including its hinterland) represented the epicentre of the waste crisis during the 1980s–90s. In this phase it survived only thanks to Regional ordinances. The two incineration plants built in the 1970s and later improved in the 1980s could treat only 40% of the water; the remainder was landfilled in very large sanitary landfill plants, created by private firms in the surrounding provinces, with strong local opposition. Increasing quantities were sent outside the region. After the last landfill site was suddenly closed in 1994, the city was forced to find a new solution. An impressive programme was launched. It foresaw in the long term the achievement of total self-sufficiency. In the medium term, the solution was provided by a large sorting plant, created and operated by a joint-venture between Amsa and four private partners, chosen among the leading operators in the field of industrial waste landfilling. Sorted waste is expected to be sent in the future to energy production (two new incineration plants are planned) and composting (with a final destination foreseen in contaminated land reclamation rather than agriculture. At the same time, an impressive separate collection programme was launched, rapidly reaching the level of 30%. Special arrangements were necessary in order to adapt the philosophy of separate collection to a congested city with limited space available both in the streets and in the houses: collective bins were put on large blocks and were allowed to be placed on the street only at certain times; plastic waste was collected in yellow sacks. A network of green compost plants was created in order to receive clean organic waste from 'large producers' and households. Around this system, a large number of small firms were created, offering specialized activities such as transport of bins to the streets during the night, collecting, sorting and pre-treating before recycling. Though with some problems in the starting phase (many irregularities were discovered) and a rising final cost for the citizen, the system seems to have now reached the maturity phase.

Two other metropolitan areas Turin and Rome, on the contrary, have exploited their own very large landfill sites, opened decades ago on what were originally quarries providing construction material during the urban boom in the 1950s and 1960s. The MSW service could concentrate on urban issues, and the problems it had to tackle were basically street cleaning, traffic, etc. Both introduced street bins in order to allow some mechanization and reduction of costs. The two big landfills are still functioning, but the end is now in sight. Though with less pressure, both cities are envisaging a similar strategy to Milan. While Turin is more oriented towards a new waste-to-energy plant receiving all waste that is not recovered through separate collection, Rome envisages a strategy based on mechanical sorting prior to energy production and composting. Again, compost is likely to be used in land reclamation.

Brescia and Modena represented during the 1970s and 1980s the most advanced systems in Italy and were models of good administration, not only in the field of MSW. They are typical medium-size towns of Northern Italy, with a population of around 200 000. They are among the pioneers of separate collection, yet this has always been considered basically as an instrument for involving the population in waste-concerned decisions and as a type of showroom of environmental commitment. The core strategy continued to be represented by hard technologies. Modena first, and now Brescia, have thus concentrated their efforts on the development of waste-to-energy projects. Separate collection finds a new justification in the need to maximize the energy content and avoid undesirable material flows and especially organic waste. It is expected to reach a diversion rate of 30% through separate collection; the remainder will be incinerated. The WTE plant in Brescia, newly in function, with a capacity of 1200 tonnes/year, is expected to receive waste from two other provinces in addition to Brescia. The city is thus trying to attract other municipalities and now sells its treatment capacity at a low price (around 50–60 Euros/ton). The low operation cost is possible due to technical solutions aimed at maximizing energy recovery, but also due to the integration with the already functioning city system of district heating.

Naples is the capital city of the Campania region, with 1 000 000 inhabitants (2 000 000 including its hinterland). In the recent past, the quality of local government both in Naples and in the Region of Campania was very poor. This was particularly true for the environmental policy of the elected bodies. Today, all the regional government members in charge of the environment during the 1980s are either on trial or in jail. In those days, waste management was a huge business for the *Camorra* (organized crime): exhausted quarries of sand and gravel were crammed with waste, for a price varying from 7–10 ECU per tonne. As controls did not exist, hazardous industrial waste was often disposed of along with the MSW in an indiscriminate way.

Until 1993, the Region of Campania did not have any specific legislation on waste, and no RWDP. At that point, the situation became unbearable: the area of Naples was called 'the dustbin of Italy' by the press. In this year, a law issued by the regional government banned private firms from running landfills. In the meantime, 18 public consortia for waste management were instituted.

With the new law, the control over landfills became effective and the environmentally worst landfills began closing down. Moreover, the remaining landfills were run by public sector bodies, either consortia or Municipalities. The manager of the consortium or the Mayor did not want to go on trial for a landfill he or she was not making profits out of. As a consequence, he or she had an incentive to run it properly. Today, the number of small, uncontrolled landfills is still decreasing, and the opening of larger, environment friendly landfills is beginning to satisfy the demands of the population, as the *Camorra* has lost much of its influence.

The Province of Padova, in the North-East of Italy, has a population of around 800 000. The Northern part of the Province is characterized by a large number of medium-small centres, in the middle of an area that has seen an extraordinary industrial development over the past 20 years. Urban sprawl, chaotic territorial development and the coexistence of rural and industrial activities characterize the territory. Until recently, MSW collected from each town was simply landfilled in a nearby plant. Recently, the increasing difficulty in opening new landfills has encouraged the creation of a consortium to foster separate collection. The consortium initially did not substitute the municipalities, but just acted as a coordinator. Increasingly, this role evolved and became more central. The consortium now sets technical standards and guidelines that are followed by each municipality, thus leading to a close integration. The strategy, devised in cooperation with a consulting company linked to the environmental association Legambiente, is extremely ambitious with respect to recycling. This strategy is surprising in an area that is definitely not famous for its 'environmentalism', but is understandable given that the area is too small to afford a WTE plant, and given that there is fierce resistance to the possibility of being included in some larger area sharing the provincial capital's problems; while mechanical sorting is perceived as unreliable. A very demanding recycling target has been set (70%), and already achieved in at least a significant part of the area. The overall performance is above 50%. The remaining waste, mostly inert material, is landfilled. Door-to door schemes for all recyclables, separate collection of green and organic waste, incentives for self-treatment (eg. home composting), *dechetteries* aimed at diverting flows of potentially hazardous waste, are behind this success. Operation is increasingly delegated to private firms under very strict performance requirements. The consortium sets technical regulations for collection and directly coordinates the supply of activities such as sorting plants and pre-treatment, again with a number of arrangements and contracts with private suppliers. All private operators involved are small local firms, specialized in particular activities, since all organizational and project work is carried out directly by the consortium and its consultants.

Rimini is one of the most important tourist sites in Italy, and perhaps in Europe. The resident population is around 120 000, but during the summer season, its population can reach over a million. Given the high number of concentrated organic waste producers (hotels, restaurants, etc.) the chosen strategy, now under implementation, puts great emphasis on composting. Organic waste is collected separately and sent to dedicated plants for green compost. The efforts of the local municipal firm, Amia, have been concentrated on the search for effective ways to dispose of compost and develop marketable products out of it. Considerable research is carried out, in cooperation with scientific institutions, in order to single out product flows with characteristics that are suitable for specialized agricultural uses, with the aim of extending the compost technique to other, perhaps less clean, organic waste flows (e.g. sewage sludge). At the same time, an attempt to improve separate collection rates for other materials is based, among other things, on an experimental use of bonus prices.

Perugia is a medium-sized town in Central Italy, which in the 1980s carried out one of the first schemes based on mechanical sorting and end-of-pipe recycling. The operator is Gesenu, a mixed joint-venture between the municipality and Sorain Cecchini, the same firm operating the large landfill of Rome. The results of the plant were quite poor in the first phase: neither compost nor Rdf were ever sold on the market. Instead, Gesenu developed a strategic plan aiming at disposing of the plant's output in small landfills, with a very rapid turnover. The treatment process, though not producing marketable products, at least creates inert waste that can be landfilled without too many problems and with low aftercare costs, thereby allowing a rapid rehabilitation after use. This strategy, refined over time, can be seen as an anticipation of what would later happen in Milan and is likely to happen in Rome. Significantly, Sorain Cecchini is a partner in the company opearting the sorting plant in Milan and is likely to enter into a similar agreement in Rome. The parallel development of separate collection and the creation of dedicated composting lines for green waste are now complementing the waste management system.

Financial structure of MSW management

The MSW service is traditionally financed by a local tax raised on the basis of the size and type of property. It is up to the LA to decide in many cases whether to bring certain categories of waste into the public service or not. In the first case, producers are treated as if they were households: they may use the service according to municipal regulations (which may also involve the obligation to deal with certain waste separately) and join the taxation system. In the second case, waste producers are treated as private polluters and have to make their own arrangements.

The MSW tax has to cover all the expenses incurred by the LA for the provision of the service, without any profit. However, this system can barely guarantee an adequate volume of financial resources:

- Municipalities have adopted this system only partially. In many cases, tax revenue covers only a small fraction of the cost, while in other cases, especially in the case of smaller municipalities, MSW taxes represent a source of income which is used to subsidise other activities.
- In the absence of appropriate accounting (which occurs for instance in the case of direct management), the evaluation of production costs is highly problematic, particularly with respect to capital costs and general costs.
- The mechanism does not generally allow for depreciation costs, and when it does, it regards past investment and not new investment. In a highly dynamic context, with significant capital expenditure, this feature curbs the investment capacity of municipalities and creates strong pressure to privatize operating firms.
- In the context of MSW operation - and especially when a municipal firm has been set up - many other activities which are different from waste are considered as well: from street cleaning to removal of snow.

The survey summarized in Table 12 shows a total revenue from the waste tax of around 2.8 billion Euros, in contrast to a total cost estimated at nearly 4.2

Table 12. Waste disposal fees and related costs in 1994 (million Euros)

Area	Total (million Euros) Tax	Total (million Euros) Cost	Per capita (Euros) Tax	Per capita (Euros) Cost	Coverage
North West	896	1281	61	87	70%
North East	519	784	51	76	66%
Centre	620	744	57	69	83%
South	514	994	38	74	52%
Islands	186	371	32	63	50%
Italy	2734	4174	50	76	66%

Source: Ministero dell'Ambiente (1997)

Table 13. Capital expenditure funded by the general budget for waste disposal (million Euros, 1986–1995)

	1986	1987	1988	1989	1990	1991	1992	1993	1994	1995	Total	Average
Granted												
State	–	16	181	73	100	43	39	42	90	82	666	67
Regions	–	–	542	462	514	431	423	406	388	369	3535	354
FIO	151	141	133	256	–	–	–	–	–	–	681	68
Total	151	156	856	792	614	474	462	449	478	452	4882	488
Payments												
State	–	–	–	2	41	46	63	47	43	49	292	29
Regions	–	–	67	139	161	128	–	–	–	–	495	49
FIO	–	–	–	–	–	–	–	–	–	–	–	–
Total	–	–	67	141	202	174	63	47	43	49	787	79

All values are constant at 1995 prices
Since no data was available for Regional expenditure after 1991, the average of previous years has been adopted
Source: Our elaboration on Ministry of the Environment (1997)

billion euros. The average coverage rate is 66%, somewhat higher in the North West and the Centre. These data should be evaluated with care for the reasons given above.

Table 13 shows the evolution of subsidies granted by the National government, Regions and the special programme FIO[9] throughout the decade 1986–1995. On average, nearly 490 million Euros was spent every year, forming a total of 4900 million. At that time, only a small fraction (16%) of this figure had actually been spent, yet many projects were being planned.

Dlgs 22/97 introduces into this picture a substantial shift, towards a more radical and proper implementation of the polluter pays principle. Further novelties have been introduced, or are being studied, in most Regions, which currently are adapting their own waste legislation to the new national outline law. First of all, the law makes a clear statement regarding the need to reach total cost coverage through waste disposal charges paid by final consumers.

The structure of costs and employment in the Italian MSW management system

A survey conducted by Iefe in 1992 showed that the largest part of costs (60%) was still attributable to collection and for the largest part to personnel costs (Ascari *et al.*, 1992). Only 2 years later, this situation was significantly different. Disposal costs had risen dramatically, and in many areas had duplicated in a few years. In the most congestioned areas in the North (especially Lombardia) landfill prices had reached 100–150 Euros/tonne and were still growing. Despite these developments, MSW operation remains a highly labour-intensive activity. Given the impossibility of estimating total employment rates, for the same reasons reported above, we can estimate figures in the order of 18 000 employees for MSW collection, 1000 for separate collection and recycling, 600 for incineration plants and 1700 for landfills (Cottica and Kaulard, 1996; Kaulard and Massarutto, 1997). According to the same studies, a 'greening' of MSW management options would imply a significant increase in the volume of employment and an increase in average management costs. This increase is expected to be significant but not dramatic (around 10–18%), and in any case it involves a drastic shift in distributive patterns, from rent and intermediation to salaries and profits.

The concept of costs that has been determined explicitly mentions capital expenditure and depreciation among them. This does not mean that state and regional transfers to the waste disposal sector will suddenly end, but they will play an increasingly minor role from now on. Second, the structure of the charge has been radically modified. While the previous waste tax was correlated to the size of the property, with only a limited number of possibilities for adjustments according to real waste generation potential, the new waste tarif' is explicitly considered as a price paid in exchange for a service; parameters being set as far as possible in terms of the real quantity and quality of waste generated. Great flexibility is given to municipalities, most of which are not likely to change very much of the present structure. They are encouraged, however, to adopt at least an incentive approach, aimed, for example, at rewarding certain behaviour patterns (e.g. participation in separate collection schemes, readiness to try home composting). A number of cities are already experimenting systems aimed at measuring on an individual or, at least, collective basis waste generation and separate collection results.

Landfill tax, whose total revenue is now around 90 million Euros, is tied by most Regions to the financing of Regional Environmental Agencies, which are in charge, among other duties, of all the control and monitoring actitivites, licensing procedures, etc. This prevents the use of these funds for other purposes, like the financing of new initiatives: in some Regions, such as Lombardia and Toscana, it is foreseen that an extra charge levied on waste disposed of outside the respective 'optimal basin' will serve for this purpose. At the same time, the creation of a new system for the valorization of packaging waste entails the reform of the previous financing system adopted for Obligatory Syndicates. This topic is considered in more detail below.

The new regime for packaging waste

The institution of Obligatory Syndicates for packaging material (law 415/1988)

A new strategy for managing waste flows involving producers and users

In 1988, pressure from green and environmental groups, as well as the debate at the EU level about the best possible strategy to reduce waste production, led the Italian government to issue a new piece of legislation introducing urgent measures to reduce waste production. This law consists of a number of measures representing a mediation by the Italian government, through its Ministry of the Environment, between green and environmental groups on one side and the industrial sector (divided into several factions) on the other. Plastic was undoubtedly the target of an aggressive campaign on the part of other packaging producers. As a result of the technical innovations and radical changes in the trade sector during the 1980s, plastic packaging production rose at very high rates, worryng both the environmental groups and the other material packaging producers. The former felt uneasy with a material that encouraged wasteful attitudes on the part of the consumer (not reusable, not recyclable, difficult to dispose of, etc.), the latter (glass producers in particular) were faced with a situation where their share of their market was disappearing in favour of the new materials (PVC and PET bottles, paper/plastic/aluminium composite containers, etc.).

One of the measures introduced by the 475/1988 law was a tax on non-biodegradable plastic shopping bags amounting to 0.05 Euros per unit. The aim of this tax was to give an economic incentive to multi-use shopping bags. The results were miserable. The incentive was too low to encourage virtuous behaviour on the part of the citizen. Moreover, in the space of a couple of months, 90% of the shopping bags turned out to be made of 'biodegradable plastic', as attested by the label printed on them. A ministerial scientific commission with the responsibility of checking the real biodegradability of these shopping bags has never been constituted. More sophisticated, at least apparently, than a simple tax on unit product, was the constitution (art, 9-quater of law 475/1988) of the Obligatory Syndicates (Consortia) for the recycling of glass, metals and plastic containers for liquids. Municipalities were responsible for the separate collection of these materials, but the consortia were responsible for the reaching of the ambitious targets set by the law (Table 14).

A tax on packaging units varying from 0.01 to 0.05 Euros, according to the packaging volume, was set for those packaging materials not achieving the assigned recycling target before the deadline (1992), unless included in a multi-

Table 14. Recycling targets of law 475/1988

	Minimum recycling target for the years 1990–1992
Glass	50%
Metal	50%
Plastic*	40%
Compounded	40%

*A maximum of 50%of the target can be attributed to energy recovery.

The target applies if the packaging is less than 65% biodegradable.

use scheme. Producers of packaging in the raw materials sector had been forced to subscribe to the Obligatory Syndicates. Packaging users and the distribution sector had no financial burden, but were asked to elect a representative. The legislator, on the basis of a restricted version of the polluter pays principle, chose to put the burden of packaging recovery entirely on the broad shoulders of packaging material producers. On one hand, this decision is open to criticism, since material producers are only marginally interested in and influenced by environmental pressure. They are involved in the war over the substitution of materials, which is also fought on the environment battlefield. This is the case of glass packaging producers in relation to plastic producers, with the former emphasizing the environmental performance of their material both in terms of capability to be recycled and energy saving, and the latter stressing the material saving due to the lightness and easy transportability of plastic packaging. However, they are less sensitive to the wishes of final consumers and environmental groups if compared to the packaging user sector and, even more, the distribution sector. And, in another sense, they have limited possibilities in terms of putting pressure on user and distribution sectors.

There is another important implication in giving the responsibility for the recycling of materials to producers of raw materials. The packaging material production sector is the most concentrated of the *filiére*. The idea is that it would be easier for a few firms, with a tradition of cooperation, to find an agreement. In Italy 90% of plastic packaging materials are produced by three firms, 80% of glass by eleven, 100% of aluminium by five. An important aspect of the law 475/1988 is that it is much more precise in defining the crucial element of the financial contribution of the plastic producers to their Obligatory Syndicates than is the case with the glass and metal producers. While the glass consortium and the metal consortium should be financed by the revenue derived from secondary materials and by the contribution (unspecified) of the firms joining the Consortia, the recycling activity of the plastic Obligatory Syndicate should be financed by the revenue derived from secondary materials and by a contribution equal to 10% of the price stated on the invoices of the material sold to produce plastic packaging. This different treatment afforded to the plastic Obligatory Syndicate is due to the awareness that a market for used plastic bottles does not exist, while the opposite is true for glass bottles and aluminium cans. It was expected that to reach the target set by the law, the Obligatory Syndicate for plastic would need a large amount of financial resources to start a totally new, artificial market. Glass and aluminium can Obligatory Syndicates had to mostly reorganize and develop something which already exists (separate glass collection started in Italy in 1976), taking a middleman role between the municipalities and the market for secondary raw materials.

It is interesting to note that the law 475/1988 only marginally considered packaging reuse. With the exception of the environmental associations, the other actors seem to agree in considering packaging reuse an impractical way to obtain waste avoidance. Even the glass packaging subsector, the one that probably would gain a share of the market from strong legislation in favour of packaging reuse, prefers recycling to reuse. One reason is to be found in the pressure from the big food and drink producers, intent on marketing their own brand image and unwilling to use standardized glass containers, which would make glass bottle reuse more economical. Another reason could be the risks involved: plastic packaging producers can produce PET reusable bottles (as they already do in Belgium) which are cheaper, safer and give better transport fuel economy than glass bottles.

The Obligatory Syndicate for glass packaging recycling

As already mentioned, the separate collection of glass bottles among householders and the subsequent treatment and recycling of this packaging material is an activity that already had a long tradition in 1988, when the Consortium was constituted. The objective of the Consortium is to improve and expand the recycling of glass through conventions with municipalities, incorporating pre-existing structures. Currently more than 80% of the Italian population is served by a separate glass collection scheme. Most of the glass is collected through bottle banks. Although internal collection is constantly increasing, imported glass cullets still represent 10% of the total glass recycled in Italy (Figure 3).

The members of the Consortium are the 11 producers of glass bottles and containers and the approximately 500 firms which operate in the collection and processing stages. The annual budget of the Consortium is about 400 000 Euros and is financed by the material producers only, while the total turnover of the glass recycling sector is estimated to be not less than 40 million Euros. Glass cullet prices in the secondary raw material market have been relatively stable. Only recently (since March 1997), as a consequence of the dramatic decrease in raw material prices, has the market for cullets been experiencing a recessive phase (Figure 4).

The Obligatory Syndicate for aluminium packaging for liquids, COALA-RAIL

This Consortium has many similarities with the one for glass: it has a streamlined structure with a yearly budget of 250 000 Euros; it pre-existed the

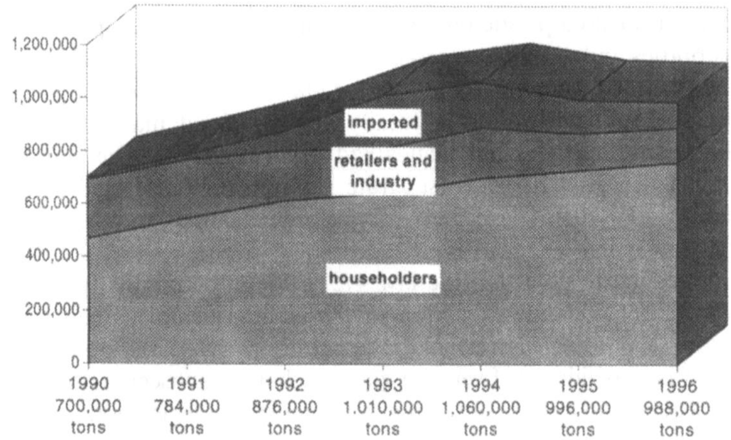

Figure 3. Glass cullets collected in Italy according to its source, tonnes. Source: CoRiVe

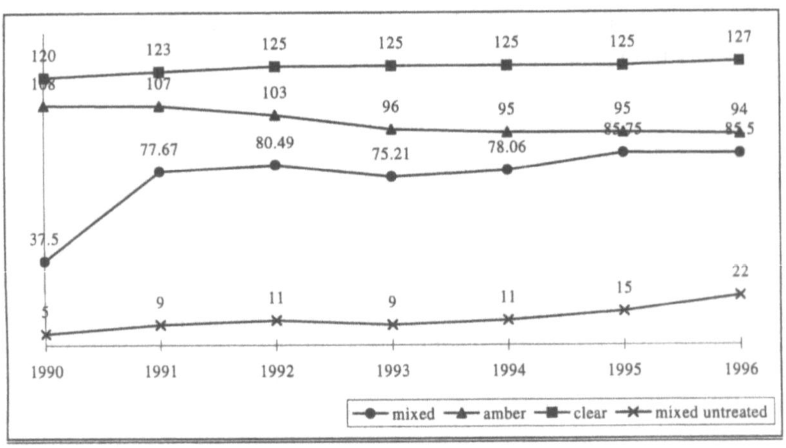

Figure 4. Secondary market prices for glass cullets, Euros/tonne. Source: CoRiVe

law 475/1988 in the form of a voluntary organization (RAIL) and, finally, as in the case of glass, there is also a secondary raw material market for the material collected through this Consortium.

RAIL is a voluntary organization founded in 1985 by the big producers of aluminium, Alcan, Alcoa, Pechiney, Reynolds and Vaw, with the objective of expanding aluminium recycling (Figure 5). With more than 400 000 tonnes/year, Italy is the biggest European producer of secondary aluminium after

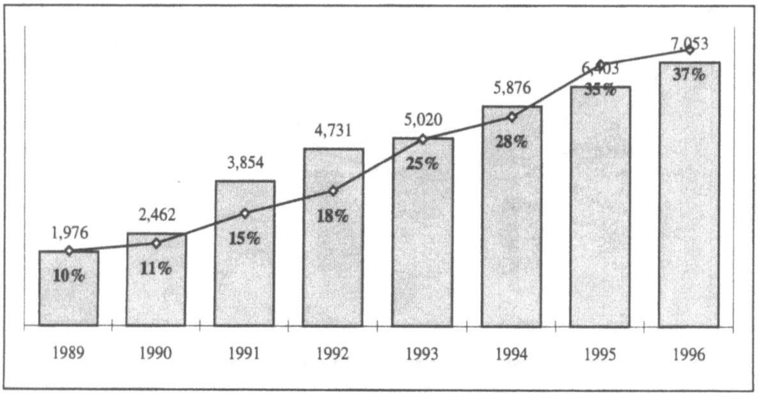

Figure 5. Aluminium cans collected in Italy and its diversion rate, tonnes/year. Source: Coala-Rail

Germany. To satisfy its production capacity, about one third of secondary aluminium has to be imported (mostly from Spain and the other countries of the Mediterranean basin).

Since the type of aluminium, 'can stock', used to produce beverage cans, is not produced in Italy, COALA, the Obligatory Syndicate for aluminium instituted by the law 475/1988, was inactive until 1992, when RAIL, through a private contract, effectively incorporated COALA and its institutional objectives (Figure 6).

As in the case of the Obligatory Syndicate for glass, COALA-RAIL's mission is to improve and expand the recycling of aluminium cans, functioning as a middleman between municipalities, voluntary associations, other institutions and the approximately 9000 aluminium recyclers.

The Obligatory Syndicate for plastic packaging for liquids, REPLASTIC

As already mentioned, the Obligatory Syndicate for plastic packaging for liquids is very different from the other two instituted by the law 475/1988. The aim of Replastic is to promote the separate collection of plastic containers for liquid products, both for food (beverages, oil) and non food (mainly detergents) products, and to recycle them.[10] For this material, Replastic has to organize a separate collection system in conjunction with the municipalities, and independently organize the processing of the collected materials transforming them into marketable products.

The participation in the Consortium was compulsory for both producers and importers of raw material used for bottles (plastic granulate). That meant the involvement of a very small number of powerful subjects. These were three multinational corporations: Solvay-Dow Chemicals, Eni and Montedison.

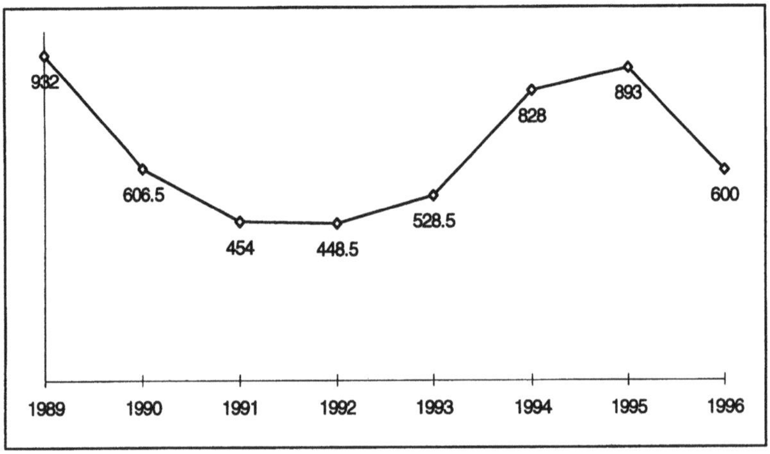

Figure 6. Secondary market prices for aluminium, Euros/tonne
Source: Coala-Rail

After a pause for reflection, the Obligatory Syndicate (OS) for plastic packaging for liquids was established on July 1st, 1991. By clearly specifying a 10% contribution for whatever kind of plastic used for plastic containers for liquids, the law did not provide loopholes as in the case of glass and aluminium. The main feature of the Consortium has been the large amount of this contribution. A contribution of 10% on raw materials meant, in 1991, some 17 860 000 Euros, when the quantity collected and recycled was negligible (Table 15). No comparison with the contributions to the other OSs is possible.

But Replastic's task was very difficult. Only thanks to these huge contributions was it possible to offer the municipalities economic incentives which were high enough to encourage them to start the separate collection of plastic bottles (Table 16).

The municipalities involved, after some initial diffidence, were happy to organize the separate collection of a waste fraction for which there was a guaranteed fixed price. For other recyclable waste fractions, such as paper, the high instability of the secondary material market price is responsible for the limited spread of separate collection schemes involving these waste fractions. The fixed price for the quantities of bottles collected by the municipalities was formalized through an official agreement with ANCI (the association between the Italian 'Comuni'), Federambiente and Assoambiente (the associations of the privately and publicly owned waste collection firms) (Table 17).

Replastic then helped set up a service network of collection points where municipalities could deliver collected materials to be baled to make transportation to the processing centres more convenient. This initiative also helped aggregate small municipalities: the separate collection of plastic bottles

Table 15. Transfers from plastic producers to Replastic, million Euros

1991	17.8
1992	21.8
1993	22.6
1994	30.7
1995	45.2
1996	28.5

Source: Replastic.

Table 16. Trends of separate plastic collection

	1991	1992	1993	1994	1995	1996	1997
Municipalities involved	361	601	1 018	1 550	2 133	3 197	3 334
Inhabitants (× 1000)	4 500	5 950	9 300	14 700	21 100	31 200	33 800
Quantities (tonnes/year)	2 000	4 600	11 200	19 200	33 000	62 000	

Source: Replastic.

Table 17. Prices guaranteed by Replastic for the collected materials (Euros/tonne)

	Delivery location		
Impurities content	At the transfer centre	At the sorting plant	At the regional sorting plant
<10 %	200	225	240
11-25 %	150	175	190
>25 %	Discretionary	Discretionary	Discretionary

Source: Replastic.

presents prohibitive costs for small user catchment areas. Not all the collected bottles and containers have to be materially recycled. Replastic offers an economic incentive for incineration with energy recovery. In 1996 almost 35% of the 62 164 tonnes of separately collected plastic bottles returned to the waste stream, 20% was stocked,and only 31% was recycled with material recovery.

In order to find economic ways to reuse plastic packaging waste, Replastic has devoted huge amounts of money to the research and realization of new treatment plants. One of the results of these efforts is that Italy is the European leader for PET recycling. Recycled material prices are obviously strictly correlated with raw material prices, which fell dramatically in 1996. The resulting fall in PET prices reduced the producers' contributions in 1996 by an appreciable extent. Then, after the enforcement of the new 'Ronchi decree', raw material producers decided to end their contributions to Replastic. Nevertheless, the Consortium still has funds amounting to about 40 million Euros.

The voluntary syndicate for paper and cardboard, COMIECO
In contrast to the previous OS, instituted by the law 475/1988, COMIECO is a voluntary organization. It was founded in 1985 through the initiative of some paper and cardboard producers. It primarily functions as an image-builder and lobbying force. In 1995 it was transformed into a Voluntary Consortium, representing more than 1300 firms (including Tetrapack), with an annual budget of about 500 000 Euros. In Italy about 28% of the paper consumed is recycled. However, this is not enough to cover the paper mill demand for recycled fibre used in paper and cardboard production. As a consequence, about one million tonnes of recycled fibre has to be imported every year. Paper prices in the secondary raw material market are highly unstable (Figure 7).

Some conclusions regarding the 475/1988 LAW
The outcome of this policy is difficult to assess univocally. In the case of exhausted lubricants, separate collection and recovery rates have rapidly managed to come up to almost 100%. In the case of glass, the OS has permitted the consolidation of already good results for the sector. Plastics, despite poorer results, have nonetheless shown impressive progress both in the separate collection schemes and in the development of valorisation techniques and capacity. As we have seen before, one important difference between the Italian OSs and DSD or Eco-Emballages is the centrality of material producers, while

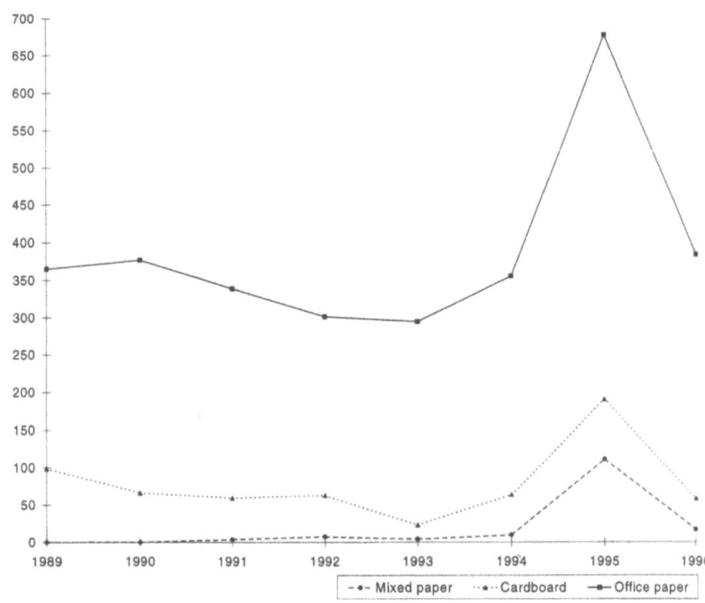

Figure 7. Secondary market prices for paper fibre, Euros/tonne.
Source: Comieco

the responsibility of the consortia is much more concentrated on final recycling than on separate collection. This difference could perhaps be seen as both the strength and the weakness of this initial phase.

OSs have concentrated their efforts on the creation of an industrial capacity for recycling and the development of recycling technologies. Paper and glass OSs have mostly worked to consolidate the already existing tradition. On the other hand, plastics have obtained remarkable results in the development of recycling technologies and in the building-up of the *filière*. Italy is now European leader in PET recycling and among the front-runners in the field of mixed plastics recycling; most of this performance is nonetheless concentrated in the segments of industrial packaging and industrial plastics waste, while in the segment of post-consumer packaging results have been far less satisfactory. In 1977, the plastics recycling sector in Italy counted 192 productive units, 2402 employees and a turnover of 385 million Euros. At the same time, however, the performance in the separate collection segment was not comparably good.

In 1996, recycling of glass bottles and containers for liquids reached a material diversion rate[11] of about 43%, recycling of aluminium cans a rate of 37% and material recycling of plastic bottles and containers for liquids a rate of about 6.5% (to which should be added a further 1.5% of bottles destined to energy recovery and an estimated 5% that is incinerated with energy recovery together with the waste mass).

None of the OSs succeeded, therefore, in reaching the target before the deadline fixed by the law (December 1992). Nevertheless, no fine was imposed, as the legislator did not exactly specify what he meant by 'minimum recycling rate equal to 50%': 50% of what? Probably of the amount of material used to produce the packaging, but since it was not clearly explained, material producers succeeded in delaying the imposition of the fines, promising to make the maximum possible effort in the future. Finally, a new law (l.649/96) was passed stating that no fine would be imposed, since the legislator was working on a new draft decree taking up the EU directive 94/62/CE, stating new recycling targets (and fines) for packaging materials. The overall performance of the OSs was therefore, not entirely satisfying. To blame is the excessively bureaucratic rigidity, the poor credibility of enforcement sanctions, and the uncertainties regarding relations with waste operators.

Unlike the German and the French experiences, OSs were still too dominated by administrative regulations and internally constrained by limited autonomy. Bureaucratic rigidity, while to some extent necessary given the absence of *filière* leaders and the great number of actors involved, curbed the entrepreneurial potential of consortia. A second reason can be found in the poor contribution of the municipalities to the functioning of the system. Law 475/88 assigned the OS the responsibility for arranging recycling facilities but not separate collection, which remained a municipal duty. Most municipalities made no effort until recently (as in the case of cities like Milan and Naples). No recycling targets were set (even if some regional plans set them) and no sanctions were foreseen for municipalities which were not committed to this policy. This was the norm wherever low cost landfills were still available and/or locally controlled. Another reason is probably to be found in the limited budgets of the OSs for glass and aluminium: their staff is composed of a very active director and not more than a couple of assistants (the same is true for Comieco). They have done a very good job, however, bearing in mind the amount of available resources.

Only Replastic had a structure proportionate to the task: 46 people. But the costs of creating a new industrial sub-sector, using as inputs the materials collected by an artificially created separate collection system, slowed all initiatives down considerably. It is likely, as stated by Replastic officials, that now that the system is working, material recycling rates will increase quickly, if the same huge amount of resources is still available. The most important lesson learned by the plastic packaging material producers from the Replastic experience is that the costs involved are very high. Therefore, there is no doubt that they will use all their lobbying power and all their ties with the government to convince regulators and environmental associations to keep energy recovery and material recycling at the same level as they are at present. The increase in business has not been affected at all by the institution of the OS. Packaging waste is increasing at a slower rate but probably due to the negative economic conjuncture and not to the activities of the OS.

Material producers seem to more or less accept the OS as a sort of taxation on their products (via the financing of OS activities): the tax burden is then transferred to packaging users and final consumers. Glass has probably gained some competitiveness with respect to plastic but this might lead to only negligible market share shifts.

As a conclusion, two main results seem to be suggested by the Italian experience of the OS:

- the absence of a 'shared responsibility principle', i.e. the lack of involvement of packaging users (fillers) and of the retail sector in the Consortia financing and activities may have had the effect of reducing the pressure from environmental associations and consumers;
- the large amount of resources available from the plastic material producers have made it possible to create a new industrial sub-sector, which, even though heavily subsidized, produces marketable materials from packaging waste. On the other hand, the recycling Consortia for glass and aluminium have not had a similar economic and technological effect.

The new system for packaging waste: Dlgs 22/1997 and CONAI

The structure of packaging waste management

The picture of Italian packaging waste management has been radically changed by decree 22, February 5th 1997, which gives full acknowledgement to the EU directive on packaging waste (EU 62/94). In short, the decree 22 sets recovery and recycling targets for packaging as a whole (and not only for liquid containers) and for each specific material[12]. The responsibility for the achievement of the targets is assigned to a new private compulsory consortium, CONAI, which, according to the shared responsibility principle, joins packaging producers, fillers and distributors (retailers).

Packaging producers and packaging users (fillers and distributors) are jointly represented on CONAI's board of directors (14 members from each category and one member appointed by the Ministries of the Environment and of Industry representing the consumer's interests) and on CONAI's executive board (eight members equally distributed between users and producers, plus the CONAI President and Vice-President); the President, elected by the partners (whose vote also jointly represents the two categories, even though the number of members from each category is different), is in charge for three years and is appointed from the opposite category to the one of his predecessor. Similarly for the Vice-President, who comes from the category opposite to the one the President represents.[13]

CONAI's internal structure represents the attempt to overcome the main difficulty of previous OSs: the lack of involvement of packaging users, who are more susceptible to environmental pressure. Now the power of the packaging material producers seems to have been curtailed, as they are sitting at the same Table with the main food and drink producers and retailers.[14]

It is estimated that about 1.5 million firms will have to join CONAI before the end of 1998. Each of them will pay a fixed entry fee of 5 Euros and an annual contribution varying according to the packaging sales (for the producer category) or to the packaging purchases (for the user category). Packaging producers and users, starting from October 1st 1998, will have to pay the *contributo ambientale*, a fee whose amount is variable according to the packaging material involved (Table 18).

The returns from this fee (collected by CONAI) are used to finance collection and sorting operations of packaging waste. It is expected that the annual budget originating from this Italian version of the 'green dot' will be around 360 million Euros.

Packaging producers and packaging users have different duties. The packaging producer has to collect and recycle its packaging from the final consumer. To fulfil these obligations the packaging producer has to choose between implementing its own system or joining a voluntary material Consortium. These Consortia are the old Obligatory Sindacates of law 475/1988 with different names: Co.Re.Pla. for plastic waste packaging (instead of Replastic), CIAL for aluminium waste packaging (formerly Coala-Rail) and CO.RE.VE for glass, and three new Consortia: CIAL (steel waste packaging) Rilegno (wood waste packaging) and Comieco (the old voluntary Consortium for paper waste packaging). These Consortia are CONAI's right hand men. They are financed directly by their members, by CONAI with 80% of the entries from the *contributo ambientale* and by the revenues from the recycled material sales.

Table 18. Recycling fees for packaging materials after October 1998 (Euros/tonne)

Steel	15
Aluminium	50
Paper/cardboard	15
Wood	2.5
Plastic	70
Glass	2.5

Source: Conai.

The Consortia are responsible for sorting and recycling secondary packaging and transport packaging waste. The responsibility for primary packaging collection is still in the hands of the municipalities, but, if they do not make enough effort, the CONAI is allowed to implement a private system of separate collection through the Consortia. Consortia activities are co-ordinated by CONAI. Packaging users' duties are to collect, free of charge, secondary and transport waste packaging, and deliver them to collection sites previously agreed upon with the packaging producers (Consortia).

The first action plan for 1998

In the spring of 1998 CONAI issued its first action plan for the following year. The action plan identifies, at least broadly, the kind of strategy that CONAI and materials consortia envisage. The starting point is shown in Table 19. As we can see, all materials are below the target level of 50–65% of recovery, though the high performance of glass, paper and wood means an average of 28% in terms of recycling, already above the minimum of 25% imposed by the packaging directive. Metals are the most distant from the target, but quantities are in this case quite small. Plastics, as we have seen, have shown very rapid progress over the past years. This performance is mainly achieved by industrial packaging. Post-consumer (primary) packaging lags behind, and is still concentrated in only a few Regions. Separate collection of household packaging, in particular, is only widespread in the North.

The target situation for 2002 is shown in Table 20. As we can see, the total quality of packaging is expected to grow by 6% (19% for plastics and 13% for aluminium). Recovery targets are fixed at the lowest level (50%), while recycling performance is planned to reach 40%. The achievement of this objective requires the fast growth of the separate collection system. In the case of metals, it needs to be set up completely; all materials need, nonetheless, a strong boost in performance.

The programme is, on the other hand, very abstract with regards to strategies and implementation measures. It seems that the preferred strategy will be to concentrate most efforts on secondary and tertiary packaging; with regards primary packaging: to improve separate collection performance starting with the lowest-cost areas, which means essentially in the North. The

Table 19. State of the art of packaging production and recycling in Italy in 1997

Material	Production (Thousand (tonnes)	Recovered (Thousand (tonnes)	(%)	Recycled (Thousand (tonnes)	(%)
Steel	428	8	2	8	2
Aluminium	62	9	14	9	14
Paper	3171	1160	37	1160	37
Wood	1061	380	36	380	36
Plastics	1685	238	14	133	8
Glass	2200	700	32	700	32
Total	8606	2495	29	2390	28

Source: Conai, 1998.

Table 20. Scenario for 2002 and five-year action plan and percentage increase with respect to 1997

Material	Production (Thousand tonnes)	Δ%	Recovery (Thousand tonnes)	Δ%	Recycling (Thousand tonnes)	Δ%
Steel	428	9	213	2563	213	2563
Aluminium	70	13	35	307	35	307
Paper	3354	6	1677	45	1510	30
Wood	1061	0	531	40	531	40
Plastics	2000	19	1000	320	360	171
Glass	2200	0	1100	57	1100	57
Total	9113	6	4556	83	3749	50

Source: CONAI, 1998

expected progress in separate collection of MSW is modest, less than 800 000 tonnes, i.e. a progression from the actual 4% to a 7% of the total MSW flow. The program fails to bind the system to any kind of target, both with respect to primary packaging and to the achievement of at least minimum levels for each geographical area.

This criticism is at the base of the Ministry of the Environment's rejection of the plan, since there is no guarantee that the packaging management system will cohere with the legal requirement to develop separate collection in all parts of the country. The internal recycling capacity is already adequate in relation to these flows, and is partially saturated at present with imported scrap materials (Osservatorio Nazionale Rifiuti 1998).

Another aspect that brought criticism from the Ministry regards the poor results envisaged on the front of packaging reduction. No reduction is foreseen in the case of any material, while plastics and aluminium expect a significant increase. Policies outlined in the plan are concentrated on the weight/volume ratio of packaging, but fail to consider a broader range of actions which include a reduction of the packaging/product ratio, an extension of the lifetime of packaging through reuse, and packaging avoidance and substitution.

Finally, the program fails to offer a proper identification of the management instruments needed. It postpones the startup of the system, with respect to primary packaging, to the creation of optimal management areas, which will cause a further delay; until then, no effort will be made to identify operational models capable of fulfilling recycling requirements, management structures, technical standards, and management agreements with MSW operators. It is also ambiguous regarding the definition of 'energy recovery', failing to provide a lower limit for energy conversion rates.

After the initial rejection of the plan, CONAI has partially acknowledged this criticism, especially with respect to the need to better specify strategies for developing separate collection and to guarantee its development in the whole country; therefore it finally gained the official approval of the Ministry. According to the agreement recently signed with ANCI (the association of municipalities) CONAI accepts to pay a fixed amount for each material in any area of the country (see Table 21); transport costs in case sorting and recycling facilities beyond a certain distance will be borne by CONAI.

Table 21. Prices paid by CONAI to municipalities for separately collected materials

Material	Price (E/tonne)
Iron	61.46
Aluminium	180.76
Paper (only packaging)	77.98
Paper (packaging and scrap paper)	12.14
Glass	30.98
Plastics	201.42
Wood	10.33

Source: CONAI, 1999

The strategy of CONAI in this phase seems to be a defensive one, still dominated by the powerful position of the materials industry, hegemonizing this sector so far, and well reflected by the plans preented by each material *filiére*. In fact the national plan presented by CONAI simply collects the individual plans of each material industry, with little autonomous addition. It is evident that the new consortium is trying to take a leading role, and that at this level a slightly more proactive attitude can be detected. In the words of CONAI's chairman, implementing the packaging directive could represent an important occasion for the Italian packaging and recycling industry, whose innovative potential is very high as well as the capacity to adapt flexibly to new market conditions. These advantages are to be sought for not in the low-cost achievement of recycling targets, but through the capacity to improve the Italian leadership in fields like plastics recycling as well as the production of environmentally friendly packaging.

It is also evident that this attempt is brought on very carefully, by avoiding any frontal counterposition. In the end the materials industry has 10 years of experience, while other actors (from fillers to retailers) do not seem enthusiastic so far to bother about packaging, and very willingly leave the initiative to the materials industry, just spending their power in the consortium in order to prevent undesired decisions. Thus, despite the cultural influence exerted by the new CONAI management, it is difficult to foresee a spontaneous evolution of the system without further stimuli from environmental policy.

Conclusion

Despite the delay, it can be seen that the Italian MSW management system has adopted a framework that is clearly consistent with the EU framework. Targets set up by the Dir. 91/156 and 92/62 are acknowledged; the approach followed for their implementation (see Table 1) is commonly regarded as a good mixture of flexibility and ambition. Dlgs 22/97 has been evaluated very positively by all the relevant actors, from environmentalists to administrators, from operators to waste producers, and this is per se an indicator of success.

Kaulard and Massarutto (1997) have shown that the intermediate model fostered by the decree is less costly in relation both to solutions that maximize energy recovery and solutions that aim at maximizing materials recovery in strict adherence to the proximity principle.

Recovery targets are quite demanding, yet available examples both in small and in large urban areas show that they are practicable – the cost is still an open question. CONAI estimates that 360 million Euros will be enough to achieve the targets, but estimates are quite rough and are heavily dependent on the type of constraints they will have with regards to primary packaging and territorial distribution of efforts.

The past history of Italian environmental policy warns us that the adherence of the legislative framework to EU parameters is not enough, and that close attention needs be devoted to the implementation phase. Targets set in 1982, for example, were only partially achieved. The achievement of new demanding objectives calls for a sharp development of the waste management system, which involves both decision procedures, planning, etc., and the organization of the industry. Therefore, the first issue to be addressed regards the compatibility of the incentive structure set up by MSW policies with the desired targets. This evaluation has at least three dimensions, which correspond to the main factors that have characterized, in all senses, the development of the MSW sector so far:

- first, the effectiveness of environmental regulations and their implementation;

- second, the incentives provided to encourage terrirorial integration and horizontal cooperation between areas;
- third, the way public utility reorganization policies such as privatization affect the capability of the system to create an efficient industrial structure through the cooperation between public and private actors.

With regards the first issue, it is positive that the traditional planning style, dominated by administrative instruments and command-without-control, is being integrated by new and more effective instruments. Roles are becoming clearer, ending the dangerous confusion between regulatory and operational functions, typical of the past, arising out of a poacher–gamekeeper game. The strategy is now more often linked to making the municipalities and private actors directly responsible, instead of concentrating these functions at the regional level through planning ordinances.

It is also promising that the public administration as a whole is showing signs of having adopted an approach oriented towards problem-solving and pragmatism, instead of a more holistic, but also more abstract and wishful, planning style. This evolution is, however, only at a very early stage. The flexibility shown with regards to the implementation of the self-sufficiency principle will probably avoid some of the problems experienced by other countries, such as Germany and the Netherlands, characterized by overcapacity, high capital costs, rigidly monopolistic local markets, while allowing, at least to a certain degree, a reasonable protection from competition. Nonetheless, the Italian regulatory system has a traditionally weak capacity in terms of orienting technological development, and for this very reason is accustomed to a rigid, formal and over-administrated regulatory process, which still discourages new initiatives.

An industrial policy is needed, with the capacity to govern the time path of technological innovation. This would be best achieved by setting a credibly stable development path, with adequate sanctions but also achievable targets, rather than through the imposition of very tight regulations and technological standards the system is not capable of achieving. Learning from the experience of other countries, Italy should perhaps develop a less formalistic, less adversarial and more cooperative framework for setting technical norms and standards.

In a way, the problem is to shift bargaining and lobbying from the implementation and enforcement level, where it is mostly concentrated at present, generally in the least transparent way, to the normative process, with the institutionalization of the stakeholders' participation in the policy making process, and, ideally, greater transparence regarding the nature of the participation.

With regards to recycling, a decisive issue will be the credibility of sanctions imposed for materials failing to meet the targets, much more than the level of the targets or administrative regulation of producers' associations.

The lesson of past policies demonstrates the inefficacy of ordinances and bureaucratic approaches. Rather, it should be noted that a strong incentive has been provided by the rise in landfill costs on the one hand, and from the development of stable market conditions for recovered materials on the other. Thus, a far more promising strategy would be the clever use of economic levers, landfill tax, incentives for recycling, together with an extension of voluntary and negotiated agreements with the recycling industry.

The second dimension to be analyzed is territorial integration. It is clear to everybody that the implementation of MSW policies requires an enlargement of management areas, in order to achieve economies of scale, encourage sufficient willingness to pay, and reach a territorial size that is reasonably consistent with the self-sufficiency principle. So far, this has proved to be a painful issue: cooperation among neighbouring municipalities has only sometimes occurred spontaneously, when a common problem has been perceived. In some cases, regional planning has encouraged this process by financing tre construction of treatment facilities or promoting the project, or even forcing municipalities into collective structures. In other cases, integration has been achieved thanks to policies arising from operators, mostly municipal operators enlarging their orbit to neighbouring areas. Availability of critical resources such as own disposal sites, excess capacity, tradition of cooperation, or complex exchanges involving other public utilities (water, transport, energy) is the key to explain these achievements. However, this has occurred only occasionally. In most other cases, the asymmetry of objectives and problems between neighbouring areas have proved an obstacle. Again, the attempt to force the creation of joint establishments, e.g. through the creation of mandatory syndicates, has often collided with local resistance, in the absence of a readiness to employ enforcement measures.

Bertossi and Massarutto (1997) point out that the design of dlgs 22/97 has limited chances of success if it continues to lean on the hierarchical imposition of optimal strategies, as envisaged by the law, providing for optimal management areas to be designated by the Regions. Planning policies should be concentrated, rather, on the creation of a suitable environment for spontaneous cooperation.

With regards to the third issue, it seems evident that it will be necessary for operators to play a more effective role. This will undoubtedly question traditional patterns of labour division between public administration and operators, since the latter will be more and more responsible for strategic decisions; and this might eventually give rise to a completely new model of local public utility organization, based on horizontal integration.

In fact, in many areas, like Emilia-Romagna, Lombardia and Veneto, there is an evident trend in this direction. However, without the development of a coherent and effective regulatory system, it is unlikely that municipalities will give up direct control of the operation of public utilities, thus preventing a true competitive process from occurring.

Notes

1. In this chapter all financial data are converted into Euros using an exchange rate of 1950 ItL/E.
2. This feature paved the way for quarrying industries becoming the main supplier of disposal capacity in the following years.
3. The proper setting of the boundary was never straightforward and gave rise to many abuses, a flood of subsequent acts and regulations, and contentious legal procedures.
4. The parameters of the waste tax, inherited from the previous regime, were the category of waste producer (household, commercial, craftmanship, agriculture) and building surfaces. The total amount had to be calculated according to the total cost. Yet the definition of cost was problematic, at least wherever a clear distinction between the public administration and the operating body was not made. Under direct management (representing the case in nearly half of the Italian population), it is not possible to properly isolate costs pertaining to waste from other costs, and particularly capital costs.
5. For example, soil analysis was required, the concentration of heavy metals in the soil had to be continuously monitored, absolute quantity limits had to be respected, etc.
6. This has happened for example to the enormous landfills used for the past 20 years by cities like Rome and Turin.
7. In fact, while totally illegal dumping is no longer feasible even in Regions where criminal organizations are stronger, the true illegal business is represented by the contempraneous control of toxic waste disposal facilities, legal and, therefore, authorized to issue legal certificates, and other facilities where waste might be illegally dumped.
8. An estimate is troublesome since data on separate collection did not identify the origin of materials – it was unclear whether it was household, commercial, or industrial.
9. Fondo Investimenti Occupazione was an extraordinary subsidy plan with anticyclical purposes, which financed the bulk of environmental investment during the 1980s.
10. 300 000 tonnes/year of plastic bottles are consumed in Italy, about 5.5 kg per capita, 15% of total plastic packaging and about 37% of primary plastic packaging consumed every year.
11. The diversion rate is defined as:

$$DR_x = \frac{\text{amount of material } \times \text{ recovered from the generators served}}{\text{total amount of material } \times \text{ available from generators served}}$$

12. See Table 5.
13. Despite these efforts to achieve equity, the fact that CONAI's first President comes from the packaging producers category is a clear signal of which category, at the moment, is prevailing.
14. For several years an association has been active for packaging reduction, whose partner are some of the main food and detergent producers (some big corporate firms, such as Coca Cola, Nestlè and Henkel, and several large Italian firms, such as Parmalat, which is market leader of dairy products in Italy, Granarolo, Lavazza, and Coop Italia, which is market leader in retailing); on the side of packagers, Tetrapack, is a very active partner with a tradition of environmental concern. This association has successfully lobbied the legislator and obtained more power in the new system.

References

Ascari S., T. Di Marzio and A. Massarutto, 1992, L'Igiene Urbana, FrancoAngeli, Milano.

Ausitra-Assoambiente, 1996, Impianti di trattamento dei rifiuti solidi urbani e assimilabili, Milano.

Bertossi P. and A. Massarutto, 1997, Verso l'industria dei rifiuti urbani, L'Impresa Ambiente, 4/1997.

Brusco S., P. Bertossi and A. Cottica, 1995, Mercato, cattura del regolatore e cattura del controllo, Economia e Politica Industriale.

CONAI, 1998, Programma generale di prevenzione e di gestione degli imballaggi e dei rifiuti da imballaggio, Milano (URL:http://www.conai.org).

CONAI, 1999, Accordo Anci-Conai, URL: http://www.conai.org/conai/gestimb/accordoANCI.htm

Cottica, A. and A. Kaulard, 1996, The environment and employment: should waste policies encourage more labour intensive options? European Commission, Dg 12.

Iefe, Eco and Eco, 1997, Per una nuova politica industriale nel settore dell'igiene urbana in Italia, Iefe, Università Bocconi, Milano.

Kaulard A. and A. Massarutto, 1997, La gestione integrata dei rifiuti urbani: analisi dei costi industriali, FrancoAngeli, Milano.

Legambiente, 1996, Ambiente Italia 1996, Roma.

Legambiente, 1997, Ambiente Italia 1997, Roma.

Lombardini S. and R. Malaman, 1994, Economia e gestione dei rifiuti, il Mulino, Bologna.

Massarutto A., 1998, L'igiene urbana, in Fondazione Rosselli, Rapporto sui servizi pubblici in Italia, Milano (forthcoming).

Ministero dell'Ambiente, 1992, Relazione sullo stato dell'ambiente, IPZS, Roma.

Ministero dell'Ambiente, 1997, Relazione sullo stato dell'ambiente, IPZS, Roma.

Osservatorio Nazionale sui Rifiuti (1998), Osservazioni sul programma generale di prevenzione e di gestione degli imballaggi e dei rifiuti da imballaggio, Ministero dell'Ambiente, Roma.

Osservatorio Nazionale Rifiuti, 1999, Secondo rapporto sugli imballaggi e sui rifiuti da imballaggio, Agenzia Nazionale per l'Ambiente, Roma.

Quadrio Curzio A., L. Prosperetti and L. Zoboli (Eds.), 1995, I rifiuti urbani, il Mulino, Bologna.

CHAPTER 5

Municipal solid waste management in Greece

A.D. Andreadakis, Y. Razis, K. Hadjibiros and D.G. Christoulas

Introduction

Since the mid 1980s, the Greek waste regime has undergone major changes instigated mainly by the directives and structural funds of the European Union. Present economic resources allocated for waste management have reached an all time high of approximately 1 billion ECU. New waste treatment facilities are emerging, the uncontrolled dumping areas are being closed down, recycling programs are being organized, and numerous sanitary landfills constructed. This procedure, however, is not devoid of problems and drawbacks. The most characteristic case is the waste management plan for Athens, which has been under discussion and continuous revision over the past twenty years.

Harmonization with the Packaging and Packaging Waste Directive (PPWD) 94/62, has not as yet been completed but it is hoped that compliance with the targets stipulated therein will help Greece alleviate waste management problems even where packaging waste is not the central issue. Until now Greece has depended and indeed continues to depend totally on landfills for the disposal of the waste, while only an estimated 8% of the total Municipal Solid Waste (MSW) is directed into material recycling. The general waste management plan which is applied in Athens – collection of the waste, composting, RDF production, landfill – has been criticized mainly on the basis that the marketability of the expected products is questionable. Perhaps the most alarming fact, however, is the lack of an effective and constructive dialogue between the key actors involved and the public, in spite of the chronic nature of the related issues.

N. Buclet and O. Godard (eds.), Municipal Waste Management in Europe, 171–202.

The development of municipal solids waste management in Greece prior to the packaging and packaging waste directive

Legislation and actors involved

The legislative framework

Greece joined the EEC in 1980, and therefore this year represents a turning point in the evolution of the Greek legislative framework in general as well as more specifically on issues concerning the environment. Consequently, the evolution of the legislation for waste during the period prior to the adoption of the Packaging and Packaging Waste Directive (PPWD) 94/62 may be divided into two sub-periods: one before and one after 1980. The second sub-period (after 1980) is also characterized by the need to harmonize Greek legislation to the EU directives and the fact that EU funds helped Greece to implement numerous projects regarding wastewater and MSW treatment.

Post-war–mid-1980s

The major concern of the waste legislation during this period was the protection of urban health. Waste was viewed as an undesirable product, which had to be eliminated in order to safeguard public health.

In 1964, the first piece of legislation on the collection and disposal of waste was passed in the form of The Ministerial Decree of 1964 (E1β/301/1964, volume II, FEK – The Gazette of the Government – 63B/1964) on the collection and disposal of wastes, focusing almost exclusively on the prevention of direct health hazards. However, during the 1970s most of the disposal sites did not comply with the technical measures included in the decree and even now, almost 35 years later, many of them still fail to do so. The reason for this may be that during the 1970s the waste generated did not present an obvious problem and the old type landfills or uncontrolled dumping sites were acceptable. Public authorities were very rarely concerned with the possible adverse effects the practised methods could have on the environment. Waste management was and still is implemented locally devoid of a general framework and planning. Waste collection and management was limited mainly to urban areas with little provision for most of the rural population.

Towards the end of this period, waste management problems became more acute due to social changes including urbanization, development of the tourist industry and changes in consumer habits as well as the rapid increase of packaged goods.

Mid-1980s to today

In 1986, law 1650 concerning ‘the protection of the environment’ was adopted creating a general framework for all environmental issues. Because it is a framework law, it is rather abstract, outlining the general principles to be followed and relying on future, specific for each particular case, decrees and ordinances for a more concrete implementation. These decrees and ordinances are still to a large extent absent or when issued not always properly implemented. Regarding waste, the following Ministerial Decrees, mainly in the form of harmonization with EU directives, have been issued.

- Ministerial Decree 72751/85 (FEK 665fi/85) – harmonization with the EC directives 78/319 and 76/403 on toxic and hazardous waste.
- Ministerial Decree 71560/85 (FEK 665fi/85) – harmonization with the EC directive 75/439 on used lubricants.
- Ministerial Decree 49541/86 (FEK 444 fi/86) – harmonization with the EC directive 75/442 on waste.
- Ministerial Decree 31784/90 (FEK 251fi/90) – harmonization with the EC directive 85/339 on containers for liquids for human consumption.
- Ministerial Decree 69269/90 (FEK 678fi/90) on the environmental impacts assessments of polluting activities.
- Ministerial Decree 82805/2224/93 (FEK 699fi/93) – harmonization with the EC directive 89/119 on the prevention of air pollution from new waste incinerators.
- Ministerial Decree 73537/1438/95 – harmonization with the EC directive 91/689 on the management of batteries.
- Ministerial Decree 69728/824 (FEK 358/96) – harmonization with the EC directive 91/156 on solid waste.
- Ministerial Decree 19396 (FEK 1546/97) – harmonization with the EC directives 91/689 and the related Commission Decisions on the terms and measures for the management of hazardous waste.
- Ministerial Decrees 114216 and 113944 (FEK 1016/97) – technical standards and measures for the waste management and the national planning.

The only Ministerial Decree explicitly related to packaging waste is the one on the containers for liquids for human consumption (B1784/90). Specific provisions with respect to packaging are also made in the framework law 1650, where in article 13 regarding packaging and environmental impacts from products it is stated that:

The use of certain materials which 'are impossible or extremely hard to comply with the general principles of prevention and especially recycling' is forbidden.

A tax will be introduced 'on enterprises that produce or import products which cause special problems in waste management or recycling, in order to cover the expenses for establishing certain projects by the local authorities for the protection of the environment'.

Although these provisions have never been enforced, they acted as the first triggering forces for the packaging chain industry.

The main actors

The actors involved in waste management in Greece during the pre-PPWD period can be divided into two groups: the responsible authorities according to the legislation (local and regional authorities and the government) and the real-life actors (waste contractors, industry, the European Union).

Local authorities

Municipalities and communities or various inter-municipal co-operation actors (associations, joined operating schemes etc.) organize the local schemes for the collection, processing and disposal of the MSW based on the local conditions. These activities are financed by local taxes imposed by the municipalities.

Regional authorities

The role of the regional authorities is limited to control and supervision of the ongoing activities without significantly affecting waste management policies. Only in the two largest Greek cities (Athens and Thessaloniki) have the regional authorities been actively involved in the planning of new waste management systems.

National authorities

Although, regional and local authorities have the freedom to apply appropriate local management systems, the national 'strategy' adopted by the government (mainly by the Ministry of Environment and Public Works) has a very strong influence on local schemes. The term 'national strategy' can only be loosely applied as in essence it does not imply a concrete, based on systematic studies policy but rather preferences and inclinations suggested and in some cases directly or indirectly imposed by government officials. The distribution of the EU funds by the government is a powerful tool to this end.

Private waste management companies

During the 1980s large multinational waste management enterprises tried unsuccessfully to enter the Greek market by proposing new waste management systems to some municipalities. Later, a more successful approach was adopted in some cases in the form of joint ventures with existing Greek waste management contractors, who aimed initially at the existing market: manufacturing collection vehicles, offering services for bulky waste collection, trading waste collection equipment etc. This collaboration helped the waste contractors to gradually expand their activities in more complicated systems of waste collection such as transfer stations, press containers for waste haulage etc. This mode of operation is supported by the Ministerial Decree 96728/96, which encourages the co-operation between private companies and municipalities towards implementation of waste management activities.

Industry

Before the adoption of the Packaging and Packaging Waste Directive, industry was involved in waste management only indirectly by promoting materials recovery mainly for aluminium cans, glass bottles and paper. The glass, aluminium and paper industries, interested in using the usually cheaper secondary material, have formed associations and helped the Greek market substantially in establishing quite effective collection and recovery networks.

The public

People considers waste as a major environmental burden, especially in large cities such as Athens and Thessaloniki. Widespread knowledge of alternative waste management systems is limited and landfill disposal is the most familiar concept, which in principle is not seriously questioned on the grounds of environmental impact. However, attempts for implementation of specific landfill projects especially in high cost and tourist areas is often met with acute opposition due to a strong 'NIMBY' syndrome, accentuated by the negative experience from the uncontrolled dumping areas.

The green organizations

These have contributed significantly to raising public environmental concern but the suggested courses of action from this quarter are often extreme, at least by today's standards. The imposition of a 'green tax' which would promote the use of reusable packaging or the active preservation of nature and natural life are among the key issues stressed by ecologists. The Ministry of Environment and Public Works is considered to have a good line of communication with green organizations, despite occasional opposition by the latter to some infrastructure projects, under implementation. The complaint is often expressed by the Ministry that these organizations should be more positive and help the authorities to convince the local public to accept environmentally sound projects, as in the case of the proposed second landfill site in the Athens region.

The European Union

This has a far reaching effect on waste management in Greece. On the legislation level most of the recent Greek regulations concerning waste simply harmonize the national legislation with the EU directives. Moreover, the EU has been responsible for a significant improvement of the existing waste management schemes, since many relevant projects such as construction of landfills, waste transfer stations, modernisation of the collection equipment, etc are financed by EU funds. It is, therefore, the European legislation in general and the European funds which are the decisive and critical factors for the emergence of a new waste management regime in Greece, rather than the adoption and implementation of any single Directive, even as decisive as the PPWD. The impact of this specific directive is weakened even more as it is still in the process of incorporation into the Greek legislative framework.

Current management of solid wastes

Applied practices

Waste characteristics

The average rate of MSW production in Greece is between 0.80 and 1.15 kg/inhabitant/day with the higher rates in the larger cities. The estimated total MSW production in Greece is approximately 3.5 million tonnes/year (YPEHODE – Ministry of Environment and Public Works, 1995). With respect to waste composition the most systematic data available are those derived from a sampling and analysis programme, which was carried out in Athens in the mid 1980s (Figure 1). The high moisture content,[1] due mainly to the high percentage of the food waste and the low paper content, are the reasons for the relatively low energy content of the Athens MSW. Although it would be expected that waste composition would gradually converge towards N. European figures, some recent, though not systematic, studies indicate a slow convergence, if any at all.

The composition of MSW in Greece is characterized by pronounced deviations between different cities and strong seasonal fluctuations. These deviations may be attributed to intense tourism in some areas (e.g. Rhodes).

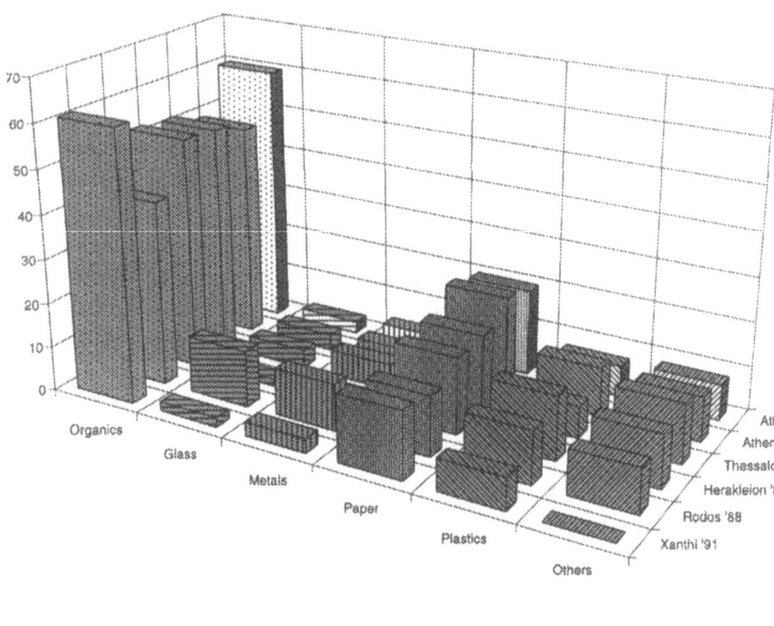

Figure 1. MSW composition from various Greek cities.
Sources: ESDKNA, 1985, 1990; SOTAMTH, 1987; Municipality of Rhodes, 1989; Municipality of Herakleion, 1981, Yiannakopoulou *et al.*, 1993

Temporary storage and collection

In most Greek cities domestic waste is put into disposable plastic bags purchased for this reason or into plastic carrier bags obtained free with purchases from supermarkets. The bags are then temporarily stored in either metal containers provided by the residents and placed near the residence entrance or in wheeled plastic or metal containers placed at strategic points in the street by the municipality for shared use. In the absence of either of the above, the plastic bags containing waste are deposited on the pavement or by the roadside.

The first step towards waste management modernization taken by almost every municipality is the use of wheeled containers for temporary storage of waste. Use of such containers serves the double purpose of safeguarding public health, especially in the summer months, and facilitating collection. The process of installing wheeled containers began in the late 1980s and is still underway in many municipalities.

Collection is carried out by trucks, usually equipped with a screw-type, high compression mechanism. These collection trucks are loaded manually, or automatically where the waste is stored in bins. Manual collection is characterized by high labour costs and disadvantages such as hygiene and work safety risks. Normally each municipality is responsible for the collection of its own waste, however in many cases the collection is carried out by an intermunicipal association for the whole regional area. In rural areas the task for waste collection is usually assigned to local contractors, equipped with an open truck. Recently the need to transfer the waste over long distances has forced some municipalities, mainly in Athens, to contract private waste management companies using more sophisticated press containers which can be towed to the final disposal area.

The successful application of a rational collection plan by local authorities depends on a number of factors. The selection of well-adjusted time-tables, the definition of fixed routes, the number of stops of the collection vehicles and the collection frequency are some of the factors which can significantly improve the technical and financial performance of the collection process. Due to several reasons (traffic congestion, poor planning/route scheduling, spatial distribution of temporary storage sites) collection processes, as practiced, are far from optimized.

Transfer and transportation

Transfer stations can be a very helpful factor in waste management due to the reduced collection and transportation costs they offer; thus making transportation distance a less restrictive parameter. In this way the selection of possible alternative landfill sites is facilitated as some fairly remote and uncontested by the public sites can become a technico-economically feasible option.

Currently, transfer stations are in a preliminary stage of development, although the need for such stations is growing due to pressures for selection of remote landfill sites. This is the case with the two largest cities (Athens and

However, even the more straightforward and less structurally demanding approach towards proper sanitary landfills met with serious problems, despite the strong backing, particularly in terms of funding, by the EU. These problems are related to the difficulties of establishing an effective 'communicative' action and reflect the existing socio-economic background of the country.

Socio-economic background

Greece, a country with deeply rooted democratic ideas, like other democratic countries of Europe and the Western world, cannot but adopt the general idea that social relations should be organized according to the principle that the validity of every norm of political consequence be made dependent on a consensus arrived at in communication free from domination. In other words, that language should have a 'communicative' use and thus the ability to co-ordinate action in a consensual or co-operative way as opposed to a forced one. However, it is well recognized that alongside this original communicative use of language aimed to reach agreement, a 'parasitic' in nature, 'manipulative' use of language aimed at bringing about compliance can be developed, still inside the domain of communicative action.

As debates on environmental issues often take the form of 'validity claims' by interested and often opposing actors to be settled through the communicative use of language, the weaknesses of this process (unwillingness to participate, strategic or manipulative use of language etc.) in a society, in this particular case the Greek society, should be recognized.

The political passions always present in the History of modern Greece and intensified during the first decade of the 20th century (1910–1920), the civil war (1947–49) and the dictatorship (1967–74) had, and to some degree continue to have, a distorting influence on the attitude of many citizens towards state affairs. The main effect of those fairly recent periods of political turbulence was the strengthening of the antagonism between political parties along the lines of traditional ideologies and doctrines. Most of the debates were and to a large extent still are focusing on 'essential' differences at the core of the opposing ideologies, with little, real attention to the merits of alternative opinions regarding 'peripheral' matters including environmental issues, which are considered to be neatly encompassed by the ideology at hand.

Important consequences of this polarization of the political scene include the following:

- Distrust on the part of the public on the objectivity of the government and a tendency to reject, often uncritically, its policies as a whole.
- Strong dependency of the local authorities from the central government and the political parties.
- Lack of systematic and timely dialogue on specific issues between government officials, the scientific community and representatives of the wider

public. Instead, a 'bargaining' approach is often adopted between officials and the affected by the issue at hand local communities.

- Discontinuity and weakening of the public administration through replacement of civil servants at high posts by subsequent governments and appointments based on party affiliations rather than competence. Such procedures magnify the NIMTO (not in my term of office) syndrome.
- Limited involvement of the private sector in the implementation of long term strategic plans due to the instability of the public administration.

The above mentioned characteristics favour a populist behaviour on behalf of the political parties and also most of the newspapers and the mass media. Not infrequently, populism and incompetence result in wrong decisions but even right decisions are often opposed by the public, as the available through the mass media information about general and specific environmental problems is poor and some times inaccurate. In general, during the last years the opinion polls have been revealing a significant shift of the citizen opinions from prejudice to a knowledgeable realism but the response of the political parties is rather slow. The so-called 'political cost' in decision making, not unknown in other politically advanced countries, still has an especially strong influence in Greek politics.

This general framework of the political life is at the root of attitudes observed in the field of waste management, including systematic opposition to local public projects of infrastructure and lack of trust in the scientific community and modern technology due to the demagogic positions of political parties, groups or influential individuals.

The case of Athens: a characteristic example

The Attica region has about 3.6 million inhabitants (1991 census), most of them living in the greater Athens area which is made up of over 50 municipalities. Recently, the total quantity of Athen's MSW was estimated at 3500 tonnes/day with a rate increase of 3.5%/year. This implies a current average production rate close to 4000 tonnes/day. Most of this quantity is disposed of to the Ano Liossia landfill and the remainder in numerous small landfills and dumping areas.

The Ano Liossia Landfill commenced operation in 1975 and 5 years later its management was undertaken by ESDKNA. It is not equipped with impermeable liners and the installed leachate and biogas collection systems are rather basic. The region is confronted with a serious problem as far as waste management is concerned, since it possesses many prerequisites for an explosive situation:

- large quantities of waste;
- a large and densely populated city with heavily congested roads;

- lack of available space at a reasonable distance from inhabited areas;
- highly competitive uses of land.

This problem has troubled authorities and public bodies (local authorities, prefecture councils, ministries, the scientific community, public institutes and political parties) for many years. The need for an immediate and practical solution with minimal public opposition has instigated several meetings, conferences, bilateral and multilateral contacts, press conferences, public actions, and parliamentary discussions over the last 20 years. In an attempt to locate suitable sites for sanitary landfills several studies involving different expertise have been assigned and a number of task groups have been appointed since 1979.

Despite this 20-year effort, very little progress has been made, and the ongoing debate has very much the characteristics of a vicious circle. This can be attributed to various factors:

- The search is limited in scope and focuses on the selection of an appropriate landfill site without due consideration of alternative schemes which are often a priori excluded for political reasons (e.g. energy recovery through incineration).
- The studies are repetitive and thus do not form a complimentary corpus, which furthers the investigation.
- The debate is clouded by a characteristic lack of good faith among the actors involved (consultants, institutes, central government, municipalities, green organizations, the public).
- The fragmentation and lack of continuity of the whole effort is enhanced by the linked changes at various levels (governmental, ministerial, administrative), which do not allow for a targeted long-term strategy.
- The NIMBY syndrome, always expected to some extent, has been raised practically to the status of sole criterion.
- The lack of progress has resulted in a continuation of the existing practices of practically uncontrolled dumping. This situation, although unacceptable from a theoretical and rhetorical point of view, in practice does not seem to pose a serious challenge to the governments and the political parties. To the majority of 'politically minded' citizens, this is still a peripheral issue unable to shake the traditional party affiliations. Thus the only serious political threat which is taken into consideration is the strong, occasionally fanatical opposition of the organized local communities in the vicinity of the selected landfill sites. The recently reported changing public attitude towards a more rational and ecological approach is limited to a rather small and low profile section of the population, especially younger people with limited influence

accentuated by their tendency to express disappointment by abstaining from public affairs.

According to the current waste management plan for the Attica region two facilities should be built, each one consisting of a composting plant and an accompanying landfill, one in the western and the other in the eastern part of the region. The first is under implementation but the latter has reached the Supreme Institutional Court of the State Council due to the reaction of the local municipalities. It seems very likely that the State's decision will be halted by the High Court on the grounds of there being insufficient and inconclusive related research on the matter.

The PPWD and the emergence of a new approach

General background

The emergence of a new approach in solid waste management in Greece can be attributed to both endogenous and exogenous factors. The main problems arising from the existing management schemes, already described, point towards a replacement of the uncontrolled dumping sites and inappropriate landfilling by properly designed and operated landfills. This is reflected by the emphasis placed in the ministerial Decree 69728/96 where it is stated that: "... national planning for the management of solid waste includes primarily the establishment of specifications and general programs referring to methods of disposal, the determination of locations suitable for disposal of solid wastes, etc, in order to avoid risks to Public Health ...", thus diminishing the importance of issues related to prevention and waste minimization as well as recovery of energy and materials.

An additional problem faced by the local municipalities is the high solid waste collection cost, which varies between 40 and 83 ECUs/tonne, constituting about 85–90% of the total management cost. The variability in collection costs arises mainly from differences in local management schemes and local conditions rather than from more general technico-economical aspects, such as the economy of scale. In any case, they are considered high and the need for optimised operations is recognized by municipal authorities who have to carry this financial burden, through municipal rates.

The main exogenous factor is the EU through funding and legislation. It is worth mentioning that over the period 1994–99 the EU is expected to invest about 650 000 000 ECU for waste management projects in Greece, an amount which represents about 70% of the total budget. The operating costs for these projects will be met through municipal rates. EU legislation is another important factor. By placing the emphasis on sustainability through minimisation, recycling and energy recovery the EU poses a new challenge. Greece, although maybe not at the desired rate, is progressively adopting the relevant EU directives and this is expected to have a significant influence on future developments and may lead to the adoption of management schemes which are not currently in the front line.

This situation is reflected by the current strategy of the Ministry of Environment and Public Works. According to the Ministry's management programme (1996), in addition to proper sanitary landfills, emphasis is also placed on the development of needed infra-structure, promotion of recycle programmes, establishment of the necessary organizational background and improvement of existing legislation. A 105 000 000 ECU fund recently allocated for projects all over the country (excluding the Attica region) has been distributed as follows: new sanitary landfills 48%, restoration of old landfills 5%, recovery projects 24%, management of hazardous wastes 8%, other projects 15%.

Although the emphasis on landfilling is obvious it should be recognized that the importance given to materials recovery and reuse is not insignificant. On the other hand, important aspects such as prevention, minimization and energy recovery are neglected.

Recycling and packaging waste

The PPWD, still in the preliminary stages of implementation and not yet incorporated in Greek legislation, though important, is not at present the main factor shaping the new developments and at this stage it is difficult to recognize its full impact. The Greek law that will harmonize the Greek legislation with the PPWD is still expected. In July 1998 the draft was notified to the European Commission in accordance with article 16 of the PPWD.

The actors in view of the PPWD

In view of the need for harmonization of Greece with the PPWD, a working committee was set up by the Ministry of Environment and Public Works in 1995. The members of the committee include representatives from the Ministry of Interior Affairs, the Ministry of Development, the Ministry of Economics, the Ministry of Agriculture, ESDKNA, Hellenic Recovery Recycling Association (HERRA), the Associations of the Materials Industries, such as paper, glass, plastics and aluminium, the Ecological Recycling Organization (green movement) and the Association of Retailers.

New actors appear on the scene, such as the packaging chain industry, the 'economic operators' as defined in the PPWD. Other actors already involved in waste management such as the government and the green movement introduce new arguments and proposals.

The Greek government considers the harmonization procedure a good opportunity to introduce a general law on recycling concerning a variety of waste, not only packaging. Consequently, the draft law includes provisions for many end-of-life or used products. The Ministry of Environment and Public Works aims to draw up an effective law, which takes into account the proposals submitted by the interested parties. In general, the Ministry of Environment and Public Works agrees with the French approach towards packaging waste management that is used as an appropriate model for the Greek law and it has contacted the French Ministry in order to obtain more information on the experience obtained in France.

The government should not be considered as a single body of absolutely clarified ideas and positions. Experts from the ministries participating in the working committee have sometimes expressed different positions on certain issues. For example, in contrast with certain sectors of the Ministry of Environment and Public Works, the Ministry of Development is more sensitive to the opposition by Greek industry to the establishment of a new public organization responsible for planning waste management policies and monitoring the implementation of the related legislation.

The 'packaging-chain' industry, which is the part of the Greek industry most likely to be affected by the PPWD, prepared itself by taking the initiative to form HERRA in 1992. The activities of HERRA helped Greek industry to formulate a credible proposal for the packaging waste issue. It is therefore not surprising that HERRA has been the major representative of Greek industry in the debate. The recovery project of HERRA, provides a working example of the application of the system in a manner approved by industry. Although the project is limited to 1.5% of the Greek population, the site has been visited by a large number of authorities (executives from municipalities, regions, ministries, etc.) who are interested not only in production results but in the management and financial aspects of the project.

The proposals of the industry are based on the Eco-Emballages paradigm. The position of the Greek industry in favour of the French model, seems to have been a key element in the final orientation of the expected Law towards the French approach.

The importance of the role played by multinational companies in transferring the trends from other markets through their new 'green' products and packaging, as well as through their internal information network to the Greek market, should not be underestimated. European industry associations, such as ERRA (European Recovery Recycling Association), EUROPEN (European Organization for Packaging and the Environment) and others, have also contributed significantly to the formulation of the view held by Greek industry.

From the beginning of the debate, the Greek industry has had no objection to undertaking a part of the packaging waste problem. The main concerns of the industrialists is that:

- Packaging should not bear any unjustified, from their point of view, economic burden, such as the high taxes proposed by the green movement.
- If packaging is decided to be the fund-raising point for recycling, then all the funds raised should be redirected into the new recovery schemes for packaging waste. Industrialists strongly recommend management of the

proposed system by industry coupled with governmental supervision for this purpose.

- No mandatory use or ban of specific types of packaging should be included in the law. The industrialists maintain that choice of materials will be influenced by harmonization costs and can thus be left at the discretion of the producer.

Most municipalities lack information regarding the ongoing harmonization procedure. The difficulty in the dissemination of information among the local authorities is one of the problems of the Greek regime, which consists of a large number of small communities (more than 5500). A new law on local authorities aimed primarily at reducing the number of the municipalities to 1000 has already been passed and is under implementation. The new regime is expected to help with waste management issues, since a pre-requisite of a well-planned, efficient, waste processing and disposal system is the implementation by a strong and effective municipality.

The representatives of ESDKNA in the working committee for the draft law, are not openly opposed to the general idea of the French model; however, implementation of the new plans for waste management in Athens and associated problems are generally given priority.

The elective actors of the municipalities and the regional authorities are usually in favour of harmonization and high technical standards in waste management treatment. This attitude, however, is not so persistent when it comes down to dealing with the practical problems involved in applying the modern methods of waste treatment detailed by the laws and decrees. It is not uncommon under such conditions for the elected body of local government to attempt transference of the 'implementation costs' (both in social and economic terms) to others, i.e. central government, other regions etc.

The Ecological Recycling Association (ERA) is a non governmental organization (NGO) representing the green movement. ERA was formed by concerned citizens and aims at organising voluntary recycling programmes, information campaigns and other public-oriented activities in order to promote environmentally friendly solutions in waste management. However, the component of the Greek green movement which deals with the MSW management is weak and its prospects for influencing decisions are poor. Nevertheless, ERA has submitted a complete proposal (ERA, 1997) for the Greek law on packaging waste, including:

- the introduction of a high 'green tax' on all end-of-life products and all packaged goods,
- the imposition of high quotas for reusable packaging in a variety of products, such as beer, wine, bottled water, alcoholic and soft drinks, juices, milk etc.

- the prohibition of PVC packaging and one-use products for serving food and beverages, plastic shopping bags etc.

Although the waste management companies and contractors do not participate in the working committee, they are quite active in the real life of waste management in Greece. Regarding harmonization with the PPWD, these companies have adopted an optimistic attitude in the belief that harmonization will offer new business opportunities; they feel that:

- The new recovery programmes will increase the demand for recycling equipment. This could prove a profitable new area of activity for domestic equipment manufacturers, possibly in co-operation with foreign companies.
- The inefficiencies of the existing waste management schemes along with the need for Greece to fulfil the PPWD targets may lead to recovery programmes organized jointly by municipalities and waste contractors.
- Waste management companies may undertake the role of the 'collective systems for the alternative management of packaging waste' (organizations like Eco-Emballages) on behalf of packaging-chain industries as defined in the draft Law. They realize, however, that the large packaging-chain industries are likely to organize such a system independently, rather than join one formed by waste contractors or other parties.

Interest has often been expressed by private contractors in the privatisation of collection and transport of waste with the claim that a more cost effective, efficient service than the local schemes already in practice could be provided; recycling is perceived as a good opportunity to forward privatisation in waste management. Given that recycling is an additional component of the existing schemes, the operation of recycling by private companies is not considered to be a threat to existing employment, therefore strong negative reaction on the part of the municipal employees is not foreseen.

The debate on the issue of harmonization with the PPWD

The debate on the issue of harmonization with the PPWD shares a lot of the arguments being raised with respect to harmonization concerning environmental, if not even wider in scope, EU legislation. The general attitude is positive as, especially in the absence of a concise national strategy, this process provides the driving force for an effective management of environmental issues. The financial support provided by the EU is an important added incentive to this end.

The general concerns voiced about the harmonization issue in the case of the PPWD as well as in other cases relate to the need for flexibility and sustainability. It is felt that EU legislation is shaped without adequate consideration of the particular needs and conditions prevailing in Greece. This to some extent is to be expected in view of the relative strengths of the member countries but the weakness of representation can also be attributed to the absence of well-defined arguments and proposals. With respect to sustainability, there is concern that the precipitated, through EU legislation and funding, changes leading to advanced management schemes may not be viable in the long term due to administrative and financial short-comings. The preparations for a transition towards a self-supported management of the emerging schemes, both at an administrative and financial level, progress rather slowly. More specifically, on the issue of the PPWD, the ongoing debate can be summarized as follows.

First, there is a strong positive attitude to recycling by the Greek (mainly urban) population. The majority of the public opinion surveys carried out by various actors (ESDKNA, HERRA, municipalities, etc.) concluded that recycling is perceived by the public as a major environmentally friendly activity. Moreover, the participation of the public in existing recovery programmes is considered satisfactory, although the inefficiencies of the programmes sometimes discourage participation. Consequently, the foundation for the successful communication of the new recovery systems exists, and it is known that one of the key factors in recycling is public acceptance and participation. However, the attitude to recycling is somewhat less positive in rural and in low-income urban areas.

Second, all the decisive actors support in one way or another the harmonization of Greece with the directive. The Greek industries, including those which are not subsidiaries of multinational companies, have been committed to contribute to the alleviation of the packaging waste problem. They believe that the harmonization of legislation between EU members is necessary for the promotion of a free market.

A more decisive involvement of the Greek Industry in the recovery systems, will reinforce the use of less expensive to recycle packaging materials which will hopefully prove more 'environmental friendly'. A second impact of the involvement of industry is that the methods and techniques, which will be followed by the new recovery systems for packaging waste, should prove effective because of industry's interest in keeping the costs low. Naturally, the government should safeguard this from happening at the expense of the other important factors(the environment, the safety of the workers etc.).

New fields in both technology and the economy will open up when the recovery and recycling systems for packaging waste are organized, and new jobs will be created. However, the new recovery systems which will be introduced during the implementation of the law, if to be successful should incorporate the existing parallel schemes.

Greece has less of a problem with packaging waste than other EU members. Packaging, as defined by the PPWD, constitutes a relatively small fraction (not more than 20%) of the total municipal waste stream, while almost 50% of the waste is food and garden waste. Therefore, meeting the targets of the PPWD is a relatively easy task especially if one considers the fact that the targets can probably be met by concentrating the efforts on the two more densely populated areas (Athens, Thessaloniki, with a population amounting to about

50% of the total population). This counterbalances the increased difficulties of applying effective methods in other parts of the country and particularly on the islands where the practical impossibility of finding local markets for secondary materials increases the operational costs due to transportation requirements. This difficulty has been recognized by the EU in giving Greece an extended period for compliance with the Directive.

It is believed that the harmonization could be a good opportunity for Greece to stimulate an 'environmentally friendly' image, which in the case of islands, for example, will act favourably as regards the tourist industry.

There is a significant market for secondary packaging materials, although, recycling rates vary significantly, depending on the material, as can be seen in Figure 2. Market forces (industries, traders, scrap dealers) are largely responsible for achievements concerning recycling and influencing rates for each material. The existing network dealing with the collection and trade of old paper and used aluminium beverage cans is the result of local industry's demand for secondary material, but the important contribution of scrap dealers and household efforts to the recycling issue, should not be underestimated.

Attempts for further development of the secondary materials market should focus on the following constraints:

- The existing collection method is a restricting factor and should therefore be improved.
- A common complaint of all industries which use secondary materials is that they do not have a constant supply of materials with standard specifications. This is a serious problem that discourages the producers to make modifications in their production lines or/and invest in new equipment in order to use or/and increase the consumption of secondary materials. For this reason, measures and incentives for promoting and stimulating the secondary materials market have to be considered.
- Regarding plastics, it is quite difficult to estimate the recycling rates, however, more than 36 000 tonnes of plastic scrap (mainly film) were regenerated in 1996, and almost half of this was imported (HERRA, 1996). Stimulating a secondary materials market, in particular for secondary plastics, has its limitations in Greece and at present it is limited to rather low value end-products (disposable bags, filaments and fibres etc.). Bearing in mind that it is practically impossible to apply state-of-the-art methods (i.e. feedstock recycling) practically in Greece, exports as well as new markets (i.e. multi-layer bottles) should be established. Currently exports of recovered paper to neighbouring countries (e.g. Bulgaria) is practised to some extent depending on price fluctuation in the local market. However, one should realise that such a small, isolated market as the Greek one will probably face greater problems than wider reaching, better established markets.

It is unlikely that the Ministry of Environment and Public Works will actually impose quotas on reusable packaging or a general packaging tax (industry's main concern) as any such measure would by strongly opposed by the Ministry of Economy; the possibility of this course of action, however, must encourage industry to undertake their part of the harmonization (recycling) costs.

A public organization established for this purpose will approve the systems on behalf of the government. It will also introduce the national plans and targets, in accordance with the provisions of directive 94/62. This organization will make the national plans in accordance with directive 94/62, approve licensing to the 'systems for the alternative management' of packaging waste which will be formed by the industry (like Eco-Emballages), and in general will monitor the implementation of the Law on behalf of the Ministry of Environment and Public Works. The real question is: how will it be financed? Apart from the economic resources coming from the General Budget, three additional duties on waste packaging have been proposed to finance the organization:

- Two licence-fees: the first to be paid by the aforementioned 'systems for the alternative management' in order to obtain approval from the Ministry of Environment and Public Works and the second will be paid by each economic operator to obtain a three-year valid certificate certifying that all the packaged goods of the operator in question comply with the law.
- Fines imposed on offenders.

Industry has cautioned that such an organization may expand beyond the necessary proportions, requiring large financial input which will be reflected in the recycling costs. Since the details of the organization and the proposed fees are to be accommodated by future Decrees, the issue is still open.

The Greek draft includes provisions for many other used (end-of-life) products such as car tyres, newsprint, electronic devices, furniture, construction and demolition waste etc. However, there are not any quantitative targets to met by the producers, since these will be a part of the future plans for each type of waste. The essential requirements for packaging are the same with the provisions of the PPWD. The law concerns all types of packaging waste (primary, secondary and tertiary). Regarding reusable packaging, a provision is made that the Ministry of Environment and Public Works may enforce future decrees that will promote reusable packaging, leaving the matter open for the time being. The reusable packaging issue reflects the competition between the German and the French systems, because the quotas for reusable packaging indicate a major difference in the philosophies adopted in the two countries. The Greek industry is strongly against any quotas for reusable packaging. Today, only beer and a few wine and mineral water brands are marketed in reusable glass bottles. The arguments of the industry against quotas for reusable packaging are more or less common with those voiced in other EU countries, and are based on both environmental and economic grounds. Reusable packaging quotas are omitted at present, despite the strong opposition be ecologists to this omission.

Implementation and prospects of the new approach

General background

As a result of the new approach to solid wastes management in Greece, several steps have been taken:

- The majority of the regional waste management plans have been completed and approved by the Ministry of Environment and Public Works.
- A large number of new sanitary landfills have been designed or are under construction.
- Modernization of local waste management schemes has accelerated.
- A number of 'second generation' recycling programs have been developed by municipalities (ESDKNA, Zakynthos, etc.).
- The first composting plant in Greece has commenced operation in Kalamata.
- The construction of a composting and RDF producing facility in Athens is underway and several other municipalities are considering similar projects.
- The first multimaterial recovery program has been developed by industry (HERRA).

Sanitary landfills

A modern sanitary landfill site is the starting point for almost all local and regional waste management plans. According to the Ministry of Environment and Public Works (personal communication, 1998) there are about 30 regional plans proposing suitable disposal sites that have been approved, while 15 modern sanitary landfills are in operation or under construction (Table 2). Data for two of the operating landfills (Zakynthos and Patra) are given in Table 3.

This massive investment, largely funded by the EU, aims at alleviating the unacceptable situation created by the uncontrolled dumping sites. With the exception of some materials which are source separated by traditionally existing operations (i.e. recycling of some paper, glass bottles and aluminium cans) in the vast majority of cases all waste is or is going to be directed to the new landfills.

This approach, the result of endogenous conditions and demands is at

Table 2. The current status landfills construction plan

City	Population (000 inhabitants)	Status of the sanitary landfill	City	Population (000 inhabitants)	Status of the sanitary landfill
Athens	3500	C	Kerkyra	40	B
Thessaloniki	1000	O	Kiato	10	C
Patra	160	O	Kilkis	15	D
Herakleion	120	O	Kozani	32	C
Larissa	120	O	Komotini	40	B
Volos	80	O	Korinthos	30	B
Alexandroupoli	40	O	Kos	15	B
Argostoli	8	C	Lamia	45	O
Arta	22	C	Lefkada	7	B
Veroia	40	B	Mytilene	25	B
Edessa	18	B	Xanthi	40	O
Zakynthos		O	Polygyros	6	B
Igoumenitsa	30	B	Pyrgos	27	C
Ioanina		D	Rhodos	45	B
Kavala		C	Samos	7	B
Karditsa		D	Serres	50	B
Katerini		B	Sparti	15	D
Chios	23	B	Chania	50	B

O: operating, C: under construction, B: bidding stage, D: under design

Table 3. Data on two Greek sanitary landfills

	Zakynthos	Patra
Budget (million ECU)	1.37	3.33
Financed by	EU structural funds	EU structural funds
Waste disposed of (tonnes/day)	11.1	68
Area (m^2)	30 000	70 000
Total capacity (m^3)	245 000	2 100 000
Lifetime (years)	15	18
Operating cost (ECU/tonne)	5.8	6.6

conflict, as already mentioned, with the general orientation of the EU as regards solid waste management and more specifically with the aims of the EU Directive on landfilling. The existing or planned recovery systems, including extensive and systematic material recycling and utilization of the biodegradable fraction of the waste (discussed in subsequent sections) are limited and the very slow progress being made in this direction creates serious doubts about the possibility of meeting the targets of the landfill directive (treatment of 25% of biodegradable wastes) in the case of Greece. Taking into consideration the strong negative attitudes towards incineration, it seems almost certain that Greece will be unable to meet the set targets, unless immediate action is taken, including:

- Provisions for utilization of the biodegradable fraction of the waste for every proposed sanitary landfill scheme.
- Investigation of appropriate localized management schemes other than landfilling, particularly in the case of islands and rural mountainous areas with transport difficulties.
- Expansion of the existing material recovery programmes in order to cover more materials and wider areas and sections of the population.
- Re-evaluation of the merits of energy recovery schemes, especially in large urban areas.
- Proper evaluation of the financial aspects of the alternative schemes, with emphasis on a realistic and overall long-term assessment of the cost of landfilling. This, being seriously underestimated during current studies, creates the illusion that landfilling is an exceptionally attractive choice from point of view of cost.

This wider in scope approach can be facilitated by the EU funding operations, provided that suitable preconditions and constraints are effected.

Recovery programmes

The recycling programme of HERRA

The recovery programme is being carried out by HERRA and five municipalities in Athens (Maroussi, Vrilissia, Melissia, Pefki, Filothei), under the guidance of the University of the Aegean. HERRA is a non-profit-making organization founded by Greek industries which produce and distribute consumer goods or/and packaging materials. The main aim of HERRA is to develop methods for alleviating the problems that go along with the environmental burden created by packaging waste. The programme collects and recovers packaging waste (aluminium and steel cans, plastic bottles, glass bottles, paper and board) and paper waste (newsprint and magazines, office paper etc.) from the MSW of the five participating municipalities.

The first stage of the implementation of the project involves informing the public. The 'communication team' of HERRA informs, door-to-door, the inhabitants about the recycling programme showing them how to separate and prepare the recyclables. They distribute a reusable blue bag for recyclables in which the householders are asked to place the targeted materials of the project. Then, the householders empty the content of the bag into the blue bins which have been placed nearby. Finally, the commingled recyclables are collected and transferred to the Material Recovery Facility (MRF) by single compartment collection vehicles (five in total).

At the MRF the recyclables are sorted along a transfer line by skilled workmen (hand sorting) with the aid of a tumbling screen and a magnet into 12 different products: four paper qualities, five plastic materials, aluminium, steel and glass. These products, are, with the exception of glass, balled and transferred to the secondary materials markets.

After 3 years of operation, the recycling programme has achieved significant results. Today, almost 150 000 inhabitants are served through a network of 2500 close-to-home recycling bins. More than 300 tonnes of recyclables are recovered from the MRF each month, with a recovery and diversion rate of 30% and 8% respectively (Razis *et al.*, 1997). The recovered materials are diverted from the landfill contributing to savings on resources, clean air and water, energy and landfill space.

The recycling programme of ESDKNA

Since March 1994, a collection programme for the recycling of paper has been in effect in 14 neighbouring municipalities in south-east Attica with the participation of the National Technical University of Athens. The programme has since been expanded to serve about 2 million people in 53 municipalities in the Greater Athens area. Around 3000 bins, 1 m^3 in volume, have been situated at a density of approximately 1 bin per 700 inhabitants, and access distance between 400 and 1000 metres. This programme involves the weekly or biweekly collection of paper, followed by transfer and direct delivery to private companies dealing with the recycling and marketing of used paper.

As part of the preparation for the project, a large advertising campaign was launched, focused on recycling and with the collaboration and financial support of the Ministry of Environment and Public Works. ESDKNA is responsible for the operation and the financing of the programme. The monthly operating expenses total 30 000 ECU. The gross cost of the programme, not taking the sales revenues into account, is about 0.66 ECU/kg of recovered paper. This cost could be reduced by increased public participation as the full capacity of the programme is 600 tonnes (without additional expenses),whereas the current recovery is 470 tonnes per month.

The collected paper is sold directly to private companies dealing with separation and sale of used paper, working on 4–6 month contracts, which are (in accordance with legislation) the result of public bidding. The diversion rate (clear reduction in the refuse flow) is approximately 1.5% and the recovery rate is 5%. The immediate plans of ESDKNA regarding recyclable materials is the extension of the programme to plastics and batteries.

Other recovery programmes

It should be noted that there are also other recovery programmes in various cities and towns of Greece, although mainly at pilot scale, of significant potential. The Association of the Municipalities in the island of Zakynthos carries out a recovery programme for paper, aluminium cans and glass containers, resulting in a 5% reduction of the waste disposed of in the new landfill site which was recently constructed. The programme serves 34 000 inhabitants and the recovered materials in 1996 were: paper 477 tonnes, aluminium 3.5 tonnes, other metals (mainly steel scrap) 133 tonnes and glass

6.2 tonnes. In 1998, a Material Recovery Facility is scheduled to begin operation, thereby expanding the range of materials targeted by the recovery programme and the potential waste reduction to almost 15%. The Municipality of Chania in Crete also carries out a paper recovery programme while many other efforts are currently under way in Thessaloniki, Volos, Kavala, Patra and Rhodes, to name but few.

The composting plant in Athens

The Association of the Municipalities in the Attica Region (ESDKNA) is responsible for the establishment of a regional waste management scheme. The plan involves the treatment of 3500 tonnes/day in at least two facilities. Each facility consists of an RDF producing and composting plant and a modern landfill for the residue. They claim that this solution is an environmentally acceptable method of municipal waste management in the area of Attica, with a significant reduction in the waste flow to disposal sites. The residue from the composting plant is expected to be lower by as much as four times the present figure.

Initially the plan proposed the implementation of three facilities which would cost 200 million ECU, 60% of which relates to the three treatment plants and the remaining 40% to the three final disposal sites. Due to public opposition, it is probable that no more than two such facilities will be realized. According to ESDKNA's estimations, the operating costs are of the order of 18.3 ECU/tonne.

Although the compost might be used as a fertiliser, there are problems with its quality and its poor reputation;[2] therefore, the final market is extremely uncertain.

According to the plan, the mechanical treatment and composting plant will produce compost for use in certain areas of agriculture (thus utilizing the high organic fraction of the waste); glass; and ferrous metals, by means of magnetic separation. Materials which may present problems in recycling, such us plastics and papers are shredded and pelletized and the pellets will be subsequently used as a partial fuel substitute (RDF) in industry (e.g. in cement factories). The waste from the plant will be to a large extent inert and is estimated to be in the region of 25–30% of the initial feed flow and will be disposed of to the local landfill. The first of these systems is expected to operate in 1999.

This plan has been criticized mainly because of the insufficient evidence as to the availability of and suitability to the compost market. Supposing that the whole food and yard waste is subjected to composting, the rate of the compost production will be close to 700 tonnes/day. The chances of the compost being accepted by farmers are very poor, particularly for compost originating from unsorted waste. Indeed, in a study conducted on the treatment and disposal of the Athens sewage sludge, the possibility of the useful disposal of MSW was investigated (Christoulas *et al.*, 1996) with disappointing conclusions. The compost which could be used in public green areas amounted to only a very small fraction of the estimated production rate. Its use in the restoration of 80

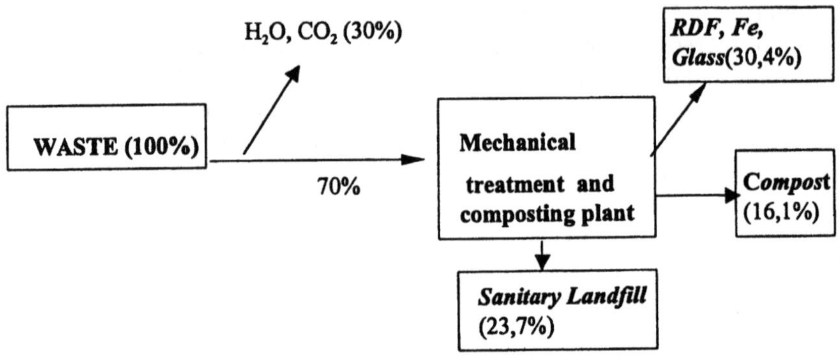

Figure 3. A simplified flow diagram of the facility

abandoned quarries in Attica was deemed questionable given the permeable nature of the limestone characteristic of these areas and high content of substances deleterious to ground water (e.g. nitrogen, organic matter etc.). Although the compost would be free from pathogens and easily biodegradable organic matter, the presence of the other substances would add considerably to the expense of such a project, not solely due to construction an maintenance of protection works but also because of the large number of sites involved. The acceptance of this method of compost disposal by the down-hill communities is very uncertain.

The composting plant in Kalamata

A composting plant for the solid waste of Kalamata commenced operation in August 1997. No operating data are available yet.

The construction firm will operate the plant for the first 18 months, after which the municipal authorities will take over. During the first two years the Ministry of Environment and Public Works will cover the operational costs of the plant, since according to a study conducted by the Technical Chamber of Greece (TEE), the revenues from the compost sales will only offset of the operating costs by between 5 and 15%.

Table 4. Data on the composting plant in Kalamata

Budget	6.67 million ECU
Financed by	EU structural funds
Waste disposed of	80 tonnes/day
Operational cost	600 000 ECU/year (at 1993 prices)
Operational cost	28.8 ECU/tonne (at 1993 prices)
Production	40 tonnes of compost/day
Production	14 440 tonnes of compost per year

Minimization and energy recovery

The Greek household waste system does not promote preventive actions along the lines of the EU policy. Incentives for minimizing the generation of wastes, by reduction at source, do not exist. First, the necessary funding for the operation of solid waste management in Greece comes from municipal rates calculated on the basis of the unit area (m^2) of the dwelling. This does not present a financial incentive, as regards reduction of wastes, or improved performance on the part of the acting authorities. It should be stressed however, that alternative methods such as the estimation of municipal rates by taking into account the number of family members or a system that would motivate the consumers to produce less wastes, for example in the fashion applied in Switzerland where the rate for each bag for storage waste is equivalent to the cost of waste management of its content, could lead to serious social and/or managerial problems.

A similar system is applied with respect to the disposal cost fixed by the managers of landfills to be met by the municipalities. Quantity is not an important parameter since the disposal cost is estimated as a flat rate (e.g. in the Attica region at 6% of the total income of the municipalities) and thus the municipalities are not motivated to reduce the quantities produced. There is also an absence of levies in order to discourage disposal of large quantities of solid wastes to landfills. The proximity principle, widely adopted, reduces the need for transportation, thus eliminating an additional incentive for waste minimization. Finally, industries are not obliged to take part in solid waste management by developing the necessary operations to guarantee the treatment and elimination of the waste generated by their products (as is the case in France through the Eco-Emballages system, or in Germany through the DSD system).

With respect to energy recovery through incineration, there is a strong negative attitude, repeatedly expressed by the Ministry of Environment and Public Works, other policy makers and peers such as mayors and local authorities (e.g. ESDKNA) and by the Green Movement.

The main arguments used to support this attitude can be summarized as follows:

- Urban areas, particularly Athens, already face a serious problem of air pollution accentuated by the frequent occurrence of temperature inversion phenomena. Therefore an additional potential source of air pollutants in the form of emissions from incineration plants should be ruled out a priori.
- The cost of incineration is considered to be very high (106 ECU/tonne according to a recent estimate by ESKDNA, 1997 for a 1200 tonnes/day facility) especially when compared with the low cost of landfilling.
- The relatively high moisture content of the solid wastes in Greece constitutes a serious drawback for the adoption of incineration.

- Energy recovery is at direct conflict with materials recovery.

These arguments carry a lot of weight and enhance the reluctance towards incineration. In view of this reluctance, it is not surprising that several attempts made to exploit the possibility of energy recovery (e.g. the 1978 Public Electricity Enterprise study for electricity generation, the mid 1980s incineration plant on the island of Zakynthos and the various proposals by international waste management companies and some municipalities) have not met with success. The only energy recovery method which is not excluded and in fact promoted by ESDKNA, with the support of the Greek cement industry, is the production of RDF. This is considered as an environmentally acceptable method, since RDF can be used as a substitute to already used fuels, especially in connection with the production of cement.

It should be noted at this point that the above mentioned arguments do not always stem from comprehensive and systematic studies. Often they are the conclusion of superficial and qualitative rather than quantitative investigations which lack the necessary detailed evaluation of environmental, technological and financial aspects. Therefore, their strength originates more from the fact that they justify preconceived inclinations rather than from scientific and techno-economical considerations. As an example, the argument that the organic content of the solids wastes is unacceptably high for incineration can be challenged on the basis of quantitative data. It has been estimated that the Athens municipal solid wastes have a net energy content of approximately 1750 kcal/g, a figure which although lower, by about 25% than the typical solid wastes in the United States (Tsobanoglous, 1993), is adequately high to support satisfactory energy production.

Prospects

Like other EU countries, Greece faces the problem of defining a new municipal waste management policy and introducing regulatory measures aiming at reducing municipal waste flows to be eliminated. At the same time the country is in a transition phase from uncontrolled dumping to properly designed and operated elimination systems, in the form of sanitary landfills. It is therefore important that the new policy and regulatory measures do not put such high demands as would block this transition from uncontrolled dumping to more reliable, hygienic and environmentally acceptable management schemes. The target of the new policy should be two-fold:

- to eliminate the existing unacceptable situation by measures which unavoidably must rely on strict command and control approaches;
- to provide incentives for consideration and implementation of management schemes other than sanitary landfilling.

The incentives and regulations aiming to serve the second target should be reasonable and realistic in terms of financial and organizational demands if a zero solution is to be avoided. The main principles of such a policy can be outlined as follows:

- Landfilling should lose part of its obvious attractiveness, mainly in financial terms, by the introduction of a landfill tax, thus making it comparable to other non-elimination management systems.
- The non-elimination management systems should be flexible in order to become realistic alternatives. The various re-use techniques (materials recovery, energy recovery, etc) should be treated without strict hierarchy.
- The targeted, alternative to landfilling, systems should bear a modest implementation cost and be feasible without major changes in the behaviour of the actors involved (e.g. the role of municipalities). Thus top of the ladder methods such as prevention and minimization or ambitious objectives for valorisation may not be easily attainable at least in the near future.
- The organizational choices to be adopted should be flexible and best suited for the fulfilment of the fixed objectives. The adoption of voluntary agreements between actors within the field of municipal solid waste management enhances this flexibility. On the other hand the often too strong importance given to the proximity and self-sufficiency principles can result in rigid frameworks, which limit the efficiency of possible management methods and should therefore, be weakened.

Notes

1. The moisture content of the MSW in Athens is estimated to 37.5% by total weight (or 60–65% by dry weight), while the average moisture content of the MSW in other European countries is not more than 20% by total weight (Ministry of Environment).
2. In France and Switzerland, the compost from unsorted household waste failed in establishing a market due to high level of glass and plastic contaminants.

References

Christoulas *et al.*, 1996, Technical, Economical and Environmental Research Study for the Management of the Sludge produced during the Operation of Phases A' and B' at Psyttalia Sewage Treatment Works. Final Report, Athens.

Ecological Recycling Association, 1997, Proposal for a law on packaging, Paper and recycling, 1 (July): 8–10, ERA and the Association of Paper-Scrap Dealers (Eds.), Athens.

ESDKNA, 1985, Ministry of Environment, Analyses of the household waste with statistically acceptable methods in the Greater Athens Region, Athens.

ESDKNA, 1990, Analysis of the MSW in Athens, Interim Report.

HERRA, 1996, Secondary materials markets in Greece, Athens.

YPEHODE – Ministry of Environment, 1995, Waste, pp. 167–180 in the Report Data, Actions and Programmes for the Protection of the Environment, Athens.

Municipality of Herakleion, 1987, Technical Report on the project 'MSW treatment: a composting facility', Herakleion, Crete, Greece.

Municipality of Rhodos, 1989, Quantitative and Qualitative analysis of the MSW of Rodos, Aristotle University of Thessaloniki, Thessaloniki.

Yiannakopoulou T., V. Aivaliotis and D. Panagiotakopoulos, 1993, Statistical analysis of the characteristics of MSW in Thrace, Greece, 3rd Conference of Environmental Science and Technology, Lesvos.

Razis Y., C. Koumenides and T.D. Lekkas, 1997, The recycling project of HERRA and five municipalities in Athens, Greece, pp. 32–40, in the proceedings of the 5th Conference on environmental science and technology, 1–4 September, Lesvos, Greece.

SOTAMTH, 1987, Quantitative and Qualitative analysis of the MSW of the Greater Thessaloniki Region, Aristotle University of Thessaloniki, Thessaloniki.

Tsobanoglous G., H. Theissen and S. Vigil, 1993, Integrated Solid Waste Management, McGraw-Hill, Inc. New York.

Ziogas Ch., 1997, Economic evaluation of the ESDKNA waste management plan for Athens, ESDKNA, Athens.

CHAPTER 6

Municipal waste management in Europe: a comparison of national regimes

N. Buclet and O. Godard[1]

A review of European waste management demonstrates that, even if not contemporaneously, each one of the five member states has encountered similar problems regarding municipal waste treatment. These problems originated in the impact of the rapid economic development experienced since the 1950s on traditional forms of organization of waste disposal. The shock resulted firstly in inadequate and saturated capacities of safe treatment, increasing environmental pressure, followed by environmental activism and NIMBY[2] forms of local reaction, protesting against even an extension of traditional techniques of disposal (landfilling, incineration). Germany and The Netherlands were the first countries to encounter such strong quantitative and qualitative modifications in waste flows that their governments decided that new concepts were necessary in the field of waste management. For a variety of reasons, including lower geographical density of human activities in France, slower economic development in a large part of Italy, and slower still in Greece, the same problems appeared later in other parts of Europe.

However, the same broad causes have not produced the same policy response in European countries, although they share a similar organization of economic integration. Available technologies and alternative approaches did not meet with the same success and development everywhere. New national regimes have appeared that present significant differences; together they produce a variegated range of answers to the same question. Are these choices quite contingent or do they respond to some rationale? Is it possible to account for such differences with the help of only a few variables?

In this chapter, the nature of the changes which occurred in national regimes will be summarized. Both technical and organizational differences will be considered in a systematic way. An account of obstacles and difficulties at the implementation stage will be followed by a discussion of implementation as a learning process, allowing for a modification of initial choices. Regimes will therefore be compared in two ways. First, key variables will be identified in an

N. Buclet and O. Godard (eds.), Municipal Waste Management in Europe, 203–224.

attempt to order the description of national regimes. The following section will consider the implementation of various policy principles. This comparative survey reveals that the key element in describing national regimes is the selective trend towards prevention adopted by some countries and not by others. Herein lies a key issue for future efforts towards harmonization at the European level.

A key element: substituting or completing policies focused on waste elimination

The development of municipal waste management in Europe has not adhered to the same timetable in different countries. Nevertheless, the five countries analysed and, more generally, the majority of European countries, are faced with the same basic problem: to what extent should a new municipal waste policy substitute or accompany a traditional regime that was in the past largely based on elimination? Except in the UK, traditional techniques of waste disposal have overwhelmingly been called into question. For a long time waste management policies had favoured technical solutions (landfilling, incineration without energy recovery) with minimal financial outlay. These policies were based on the assumption that treatment plants and landfill sites were able to absorb and solve the waste management problem definitively, without endangering human health and damaging the environment.

In the late 1980s, various factors conspired to alter the social landscape. At the local level, a general protest movement arose against elimination. The growing public concern regarding the harmful effects of landfilling, the predominant technique employed (see Table 1), fed the increasing hostility of local populations against the designation of new landfill sites. This protest movement also targeted incinerators, which were charged with emitting various atmospheric pollutants dangerous to human health (heavy metals, hydrochloric acid, dioxins). At the same time household waste systems had to cope with an increasing volume of waste that stretched treatment capacities. Financial returns and increasing economies of scale in implementing modern treatment technologies put strong pressure on traditional waste management achieved at the very local level. The need to rationalise waste management and take into account increasing difficulties in local management, but also the need for new decision-making procedures, more open to issues of acceptability (Dente *et al.*, 1998), led to a reappraisal of technological and organizational models. Techniques in use were reconsidered and the scope of waste management schemes came to include options that were either new or had been neglected in the past.

A solution frequently resorted to was to reorient national policies in order to encourage re-use, sorting and recycling, which had been neglected and in some cases virtually abandoned since the 1970s. When implementing these new strategies emphasis was placed on household packaging waste, since this type of waste revealed the highest rate of increase in industrial countries during this

Table 1. Share of landfilling in municipal waste management (1990)

Country	% of landfilling
Greece	100
Ireland	100
United Kingdom	90
Italy	80–85
Finland	78
Spain	74
Germany	70
France	52
The Netherlands	52
Belgium	49
Austria	48
Sweden	35
Denmark	15

Sources: Leppänen (1994); Royal Commission on Environmental Pollution (1993); Department of the Environment (1992); DSD (1995)

period. Packaging came to account for between 25% and 30% of the total amount of household waste by weight and nearly 50% by volume in the early 1990s. It was also the type of household waste which constituted the principal source of secondary raw materials.

Another common feature of new initiatives resided in the use of organizations producing a broad family of 'voluntary and negotiated agreements' between industrial circles, governments and local authorities: this was the case of DSD in Germany, Eco-Emballages in France, material consortia in Italy and the packaging covenant in The Netherlands (Whiston and Glachant, 1996). Such agreements are often considered as a third generation of policy instruments, taking the place that traditional regulatory and economic instruments had occupied with mitigated success (Dente, 1995). The choice of this type of approach expressed the clear will of national regulators to have firms involved in the organization of the new regime, assuming new responsibilities for an activity previously limited to local authorities.

To some extent, this shift in policy instruments was an implicit acknowledgement of the failure of authoritarian approaches in dealing with matters containing a huge economic and behavioural dimension. National regulators did not want to hurt business but intended to use decentralized choices to promote the recycling of materials and develop appropriate co-operation between actors. Basic targets and principles were still the prerogative of public authorities. In this way, organizational costs and the adverse economic side-effects of policy instruments were expected to be minimized.

Such gains in terms of acceptability and flexibility could be expected to be allied with economic efficiency only if the decentralized choices of firms and local managers dealing with waste responded to price incentives set at the right level: to provide an appropriate guide for decision-making, they had to absorb social costs generated by the use of different materials and techniques; setting up a broad-based financial organization based on undifferentiated fees was not sufficient in this regard (Defeuilley and Godard, 1997; Godard, 1999). This is certainly a serious contentious issue concerning the use of this new type of instrument in the waste management field: both DSD in Germany, Eco-Emballages in France and consortia in Italy have been attacked on economic grounds because of various incentive failures.

Differences in objectives and organizational solutions

Objectives

Beyond a shared re-orientation of waste management, national solutions differed considerably regarding the way objectives were selected, the field of implementation (the type of waste involved) and the specific requirements imposed on decentralized actors. Italy was the first European country to adopt quantitative objectives in dealing with some packaging waste: targets have been determined in Italy for a very specific category of products (beverage packaging), while France has focused on domestic packaging waste in a more general way, and Germany and The Netherlands on the whole packaging issue. In Greece the emerging tendency seems to echo the French solution. Diversity is also observable in the stringency of targets and the timetable for compliance.

All countries acknowledged the new need to developing recycling and energy recovery. Beyond that, some countries emphasized reduction at source and bottling, whilst others did not include any specific measure aimed at these goals and in fact did not seem to really wish them to be developed at all. For example, Germany and The Netherlands set up regulatory arrangements specifically limiting the use of packaging and therefore have begun to reduce the production of waste at source. Furthermore, Germany has set up a regime intended to influence the production and composition of household packaging waste flows by imposing high and differentiated charges on packaging materials. Meanwhile, Germany has not defined any clear objective regarding the overall production of packaging waste, while The Netherlands has set such a limit: the volume of packaging waste generated in year 2000 should not be larger than that in 1986. In Greece, France and Italy, reduction at source is not a set policy objective. France and Italy favour re-use, but without the inclusion of measures promoting prevention, whilst the main concern of Greece lies in the development of controlled elimination schemes, instead of habitual uncontrolled disposal of waste by the population.

In this comparative landscape we can identify three main categories of countries: those aiming at prevention plus valorisation, those only looking for valorisation, and those mainly concerned with a rationalization of elimination. Underlying these choices, there are not only objective differences in the

Table 2. Objectives in the five analysed countries

Country	Objectives	Timetable
France	Domestic packaging: 75% of valorisation[3] Municipal waste: 100% of valorisation except for ultimate waste[4]	1992–2002
Germany	Packaging: between 60 and 70% of recycling[5] Deposit refund scheme and a quota of 72% of rebottling for beverage Strict hierarchy of treatment techniques Separate collection of organic waste	1991–1998
Greece	Packaging recycling specific[6] objectives of the 62/94/CE directive: 25% of recycling	1994–2000
Italy	Beverage packaging: 50% (glass, metal) or 40% (plastic, composites) rates of recovery. Since the 1997 decree, between 50% and 65% of recovery for the whole of packaging waste (between 25 and 45% of recycling) and 35% of separate collection	1988–1992 (regularly postponed) 2003 for 1997 objectives
The Netherlands	Packaging waste volume generated in 2000 should not be bigger than in 1986. No more landfilling for combustible waste. Minimum of 60% of recycling of packaging waste. No specific goals for municipal waste.	2000

[3–6]See Notes section

problems requiring solutions and different options regarding economic development, but also different conventions determining the directions which should be explored in finding long term solutions. These conventions, encountered in countries as different as Greece and the UK, express the confidence that pollutants can be kept under control in appropriately managed landfills, without any leakage which could affect ecological health. The focus on recycling expresses a confidence that organizing new material cycles in society can overcome entropy and run as a form of perpetual movement, allowing economic growth to be based on a material throughput which only has a minor impact on the environment. The first convention conceives of post-modern society as a service society, which has succeeded in escaping from materialism and the drive for the possession of material objects. This concept of the 'dematerialization' of modern societies is well expressed in slogans such as the 'factor four' or 'factor ten' programme (von Weizsäcker *et al.*, 1997; Erkman,1998). Naturally, such fundamental differences shape the types of policy instruments being developed.

Responsibility and levels of co-ordination

In most countries, municipalities or the relevant local authorities were and are still responsible for the overall management of household waste. This is the case for France, Greece and The Netherlands. But some countries have introduced specific regimes for particular types of waste. In Germany the regulator has limited the field of competence of local authorities and has given the DSD the responsibility of managing and financing the whole chain of valorisation of packaging waste, from the collection of waste to final use after sorting. This is still an exception in Europe. In this system, industry has to bear the total cost for managing packaging waste, whilst in France, for example, industry only bears part of the cost, namely the supplementary cost generated by the separate collection and sorting of waste.

Beyond this broad distinction, several organizational models of management of waste flows and of sharing responsibilities and charges between public and private actors can be identified. Germany has chosen to divide waste flows into packaging waste and other waste and to allocate to each type of flow a distinct operational and financial responsibility, borne by different actors, even if this approach accrues additional total costs due to the multiplication of overlapping organizational schemes. Other countries have preferred not to divide operational responsibility, with only weak links between organizational and financial responsibilities, in order to retain the option of rationalising collection and recovery schemes. This last approach has been reinforced by the existence of départemental (in France), provincial (in The Netherlands), or regional (in Italy) waste management plans. These plans aim to lend coherence to various local decentralized initiatives, and ensure compliance with national regulatory objectives.

In The Netherlands, a waste management deliberation body (the AOO) has been set up to contribute to waste planning at the national level. It assumes the role of a deliberative platform for the principal actors concerned: the Ministry of the Environment, the Association of Municipalities, the Association of Provinces, environmental and consumer organizations. Whilst in other countries co-ordination is mainly achieved at the level of larger municipalities and authorities, in The Netherlands it is to be found both at local and national levels. For example, significant co-ordination has taken place between the provinces for improving the use of existing treatment capacities, involving a co-ordinated procedure in allocating waste to different facilities.

Technical and economic choices

To impose or not to impose a hierarchy of techniques?

Some countries have decided to impose a strict hierarchy of technical options for managing waste as an integral part of national policy. This is the case for The Netherlands and Germany. Others have preferred to give decentralized actors the opportunity to make their own choices within the whole spectrum of technologies, in order to find the best fit for local conditions. This is the case for France, Italy and Greece. In France, if the orientation outlined by the law of July 1992 posed a severe limit on landfilling (landfilling was restricted by 2002 to ultimate waste that could not be valorised), in order to favour valorisation, no hierarchy of valorisation techniques was established. Packaging waste management, recycling, composting and incineration with energy recovery were given equal consideration as valuable options.

Germany and The Netherlands, on the contrary, attached considerable importance to the principle of elaborating a general hierarchy of techniques as an expression of national priorities. The order of priority was the following: prevention, material re-use, material recycling, incineration with energy recovery, disposal other than landfill, landfilling. All private and local public actors had to respect this hierarchy. A lower level solution could be used only if higher level ones were not available. For example, in Germany the DSD had to reach the valorisation objectives of packaging waste entirely through recycling. The use of incineration with energy recovery was a valid option if and only if the recycling targets had already been met (DSD, 1995). Whilst this hierarchical approach might induce higher costs by imposing uniform choices in specific local conditions, it could have a significant influence on the development of new approaches and new waste management technology; specifically, it could be the launching pad for specific trajectories of technological innovation, intended to be the base of a more sustainable organization of economic activities. So, several countries have opted for a hierarchy favouring prevention and recycling, whilst others have not. Consequently, contrasting technical and industrial trajectories can be expected in Europe should such diverging conventions stabilize.

The question of establishing a national hierarchy deserves to be considered from the cost perspective. The countries that chose a sort of 'public technological interventionism', aiming to advance new technologies beyond the state of the art had to support much higher management costs than the others, initially at least. Two elements account for the higher costs of this 'technological challenge' approach in the short and medium term: first, when authorities impose one form of technology, mainly or exclusively, the management system loses the anticipated benefits of a complementary use of other solutions better adapted to some parts of the waste flow (Bertolini, 1994); this is a cost of allocative inefficiency. Secondly, regulated agents have to pay additional costs linked to R and D investments, and costs involved to develop, adapt and diffuse the new technology on an industrial scale; this cost is an investment cost which may be offset by long term returns. For example, it seems that 40% of the overall 1995 budget of DSD in Germany was used to support and develop the separate collection, sorting and recycling of plastics (Coopers and Lybrand, 1995). These two factors partly explain the cost differences observed, for example between France and Germany in the first phase of new regimes of packaging waste management (Defeuilley and Quirion, 1995). It is too early to determine the future pay-off of alternative strategies.

Financial rather than economic instruments

In addition to direct regulatory measures (targets, quotas, the banning of some techniques), most countries have introduced some form of financial mechanism in order to raise the money needed for financing actions able to fulfil their objectives. The Netherlands are an exception in this regard, since they strongly adhere to a mix of regulatory measures and voluntary agreements without accompanying financial instruments. However, the nature and level of financial mechanisms varies slightly from one country to another. Many forms of instrument are used: landfilling levies in France, in The Netherlands and in Italy, and also in other countries not specifically considered here, such as Denmark and the UK; contributions paid by firms using packaging to business-oriented organizations in charge of packaging waste management (France, Germany); levies raised on some products, such as the levy on plastic bags in Italy, which was introduced at the beginning of the 1990s and suppressed after a few years.

In most cases the rationale of these instruments is strictly financial and provides no incentive. Landfilling levies are used to finance the development of valorisation techniques but only account for part of the existing price gap between these techniques and landfilling. The contributions paid on packaging are intended to finance selective collection and sorting but may also, if cleverly designed, have a structural impact on the use of packaging (choice of materials, type of bottling technique, nature of the flow to be treated). Meanwhile, some of these financial instruments may incorporate an explicit incentive. For example, Eco-Emballages adopted, in June 1996, a progressive scale of subsidies to local authorities for recycling plastics, the unit payment increasing with the rate of sorted materials per inhabitant. This tariff was supposed to encourage local authorities to reach higher productivity in selective collection and sorting.

In effect, these instruments (levies, contributions, subsidies) could be considered as providing an efficient incentive when economic agents (firms, households, local authorities) modify their behaviour in response, harmonizing marginal costs across alternative technological options for solving the waste problem. The instruments used demonstrate some differences in this regard, but most of them have a financial dimension which does not significantly influence behaviour, disregarding cost–effectiveness issues. These are accompanying, not leading, instruments. In such cases, the main impulse comes from the regulatory constraints.

The case of The Netherlands appears to be an exception in this overview. Since the early 1980s the use of 'voluntary agreements' has been favoured by the trend towards deregulation and the disappointing assessments of command and control approaches. 'By dealing directly with the affected interest groups in society, support for a more powerful implementation was expected to increase' (Neumann, 1997). Goals were set on a sector basis through negotiated agreements. The latter took the form of common declarations of intent (covenants), including substantive targets and also a procedural dimension, between government and industry. Though the inspiration is voluntary or negotiated, what has been agreed to in the covenant is enforceable under civil law. The packaging covenant is one of the many covenants adopted in The Netherlands, as an alternative to the use of economic instruments such as charges and tradable permits.

The countries which use fiscal instruments with an incentive effect, even if this is not explicitly intended, have generally adopted objectives and targets that really require significant changes in the behaviour and choices of actors, in particular regarding technical options and production processes. This is the case of Germany, or Italy in the case of non-biodegradable plastic bags. This could have been the case for The Netherlands if they had not chosen the alternative approach of voluntary agreements. When Italy introduced a tax on non-biodegradable plastic bags, the intention was to dissuade their consumption and encourage the use of substitutes. Through the DSD, Germany took certain measures which included high contributions with rates differentiated according to materials. Rates were calibrated to cover the financing of recovery and usage of materials reaching specific recycling targets. The explicit goal was not to use incentive charges as such, but to raise enough money to cover the costs of the whole chain for each material. The combination of very ambitious objectives (recycling plus prevention) and a rigorous technological interventionism in packaging waste management explains the high level of some of the DSD fees. This indirect, less visible part, of high and differentiated contributions on behaviour, ensuring the viability of the recycling sector, stimulating a reduction of packaging flows introduced on the market, and encouraging substitutions between materials, should help to reach very constraining objectives in an efficient manner.

Table 3. The incentive content of instruments

Country	Instruments	Incentive performance
France	Levy on landfilling: 6.15 Ecu/tonne (1998) Average contribution on packaging: 0.0015 Ecu	Weak Weak
Germany	Contribution per kg of packaging: between 0.07 Ecu and 1.53 Ecu. Differentiation of 1 to 20 between glass and plastic redistribution of financial supports[8]	Strong
Greece	No instruments. Projects of adaptation of Eco-Emballages to the Greek regime	None
Italy	Contribution on plastic bags: 0.07 Ecu/unit Levy for packaging unit from 0.01 to 0.05 Ecu per packaging	Strong Weak
The Netherlands	Landfilling tax: 15 Ecu/tonne (1996) Covenant	Medium Non measurable ex-ante

The constraints introduced by the Dutch policy are different. It was not intended that the change in the behaviour of the actors, industry in particular, be induced by any financial mechanism; change should emanate from the negotiated commitment of all parties and a co-operative approach capable of creating a favourable climate for innovation and creativity in the adoption of low-cost solutions. Of course, the enforcement of the covenant could be undermined by free-riding. At present, even if all the objectives have not been attained to the same degree, there are no examples of clear failures. In the specific case of the packaging covenant, however, firms not willing to follow the covenant's prescriptions automatically place themselves under the jurisdiction of the direct transcription by The Netherlands of the European Packaging Directive. This seems to act as a credible deterrent for industry, which has managed its operations in such a way as to prevent any marked free-riding strategy.

In France and Italy, the levies on landfilling are fairly low and cannot by themselves invert the cost hierarchy between recycling and elimination. The main policy objective has to be achieved using other means. In France, the driving force of valorisation is the obligation to valorise any waste except ultimate waste by 2002. So levies on landfilling merely aim at facilitating the adoption of technical and organizational schemes for valorisation. The same reasoning also applies to packaging waste. The instruments used in France are not sufficient to incite firms using packaging for transporting and selling the goods they produce to modify their behaviour. While Eco-Emballages does support actions taken by municipalities by financing part of the costs of selective collection and sorting, the mechanism does not have any effect upstream due to the low level of contributions paid by firms to transfer the responsibility for the end of the life of the packages they introduce into economic circulation. However, the choice not to impose a national hierarchy among techniques of valorisation, and the absence of intermediary valorisation objectives for each type of technique before 2002, justified stronger incentives to ensure that targets fixed by the authorities be met in practice (Defeuilley *et al.*, 1997; Godard, 1999).

Obstacles and difficulties in implementation

Whatever the specific objectives, organizations or policy instruments chosen by individual countries or national policies have all encountered various obstacles and difficulties. In spite of real achievements, modifications were necessary during the implementation phase. What were these obstacles in the way of new waste management solutions?

In The Netherlands, the 1994 goal of reducing packaging flows in circulation was already reached in 1996 for aluminium, tinplate, glass and paper/cardboard. Not surprisingly, this was not the case for plastics. This led the actors involved in the packaging covenant to elaborate various sub-covenants, one for each material. This new subdivision gave more responsibility to the producers of specific materials, avoiding the failure of the whole covenant due to the lack of effort or difficulties experienced by some material producers. In the opposite case the government could impose strong regulatory measures on all producers, which would penalise those doing what was expected from them in the covenant. The Dutch regime also posed financial problems for some local authorities, because of a trend to under-use existing though retrofitted incineration capacities, too costly to compete freely with landfilling either in The Netherlands or in neighbouring countries. A shift in the hierarchy of technologies has resulted in a re-appraisal of the status of incineration to achieve a better use of existing capacities.

In Italy the system of Obligatory Syndicates regrouping producers for each material did not yield the expected results. For glass and paper/cardboard, such syndicates were useful in consolidating the already good recycling results, grounded in traditional practices, but they did not succeed in attaining the more ambitious objectives within the agreed time limits. The progress in plastics, though insufficient to comply with the objective of 40% recycling, is impressive, with an official 31% recycling in 1996. However, most of the plastic recycled did not belong to the category of domestic packaging waste. A serious problem for increasing the rates of recycling arose from the poor participation of the municipalities in the running of the system. The Obligatory Syndicates had the responsibility of managing the recycling of materials, but not the separate collection schemes, still in the hands of municipalities. Many municipalities made no effort at all until very recently, and no sanctions were foreseen against them for this failure. There was clearly a missing link in the recycling chain in Italy and this problem had not been overcome by the end of 1998.

In Germany, the problems encountered were not directly related to the issue of goal attainment. They mostly concerned the external issues caused by the functioning of the DSD due to an unexpected success at the separate collection and sorting stage, attributable to a very good public response. Firstly, the system faced a financial crisis in 1993–1994, because of a rapid take-off of separate collection that had to be financed and major shortfalls in the payment of contributions by packaging firms. This crisis may be viewed as a 'growth crisis'. Another problem had a more durable aspect: the lack of capacity to recycle all sorted materials. Initially the problem was side-stepped mainly by exporting huge quantities of sorted materials abroad. This 'solution' generated shocks in secondary markets abroad, with a sudden surge of low or negatively-priced materials. Since the German DSD subsidized collection and sorting operations, materials could be sold at the low price necessary to decrease stocks of unused sorted materials. Meanwhile, financial subsidies in other countries were lower (as in France) or even non-existent (as in The Netherlands), entailing ex ante higher equilibrium prices for secondary materials: recyclers had to make their living by selling these materials. Furthermore, quite often the materials collected and sold by the DSD were disposed of as 'waste' and not as a 'secondary' commodity, in order to avoid the regulations. For industry, the main problems of the DSD were not related to this export issue as such, but to its excessive overall cost for them. The lack of appropriate controls against the rent-seeking strategy developed by the waste management operators has been criticised in several quarters. Combined with the dual system of management, this led to many inefficiencies, which were still not resolved in 1998, even if the most intense period of the export of sorted materials had ended.

In France the objectives fixed in 1992 were not as ambitious as they were in Germany or in The Netherlands. However, some difficulties emerged at the very beginning of the implementation phase of Eco-Emballages. The number of municipalities committing themselves to the new valorisation remained limited, in spite of significant trials. Only a large minority of them made the appropriate investment to reach the expected valorisation rates. Financial costs involved were higher than expected, confronting municipalities with real budgetary difficulties in a period when several other environmental (water treatment, urban noise) and non-environmental issues (unemployment, urban security) demanded an increasing level of spending. Authorities were reluctant to further increase local taxes after an already sharp increase over the last 10 years. Initial rates of subsidy were reported inadequate to cover additional costs in many cases, in spite of the announced principle. In addition, the total amount of investment required for implementing national waste management legislation was greatly underestimated (several times over!) at the time this legislation was discussed by Parliament. The gap is all the wider in that, according to the départemental plans elaborated during recent years, incineration with energy recovery will become a technological fix, presumably having to treat between 65% and 70% of the total amount of household waste by 2002 (27% in 1989). Other valorisation techniques would then be condemned to a relatively limited share. Composting (8%) and recycling (19%) would together represent 27% of valorisation, versus 8% for landfilling and incineration without energy recovery (Denby-Wilkes, 1996). This huge focus on incineration does not accord with the idea that preserving decentralized choices from command and control approaches permits a diversified adaptation to local conditions. Neither is it in line with the initial objectives announced as the basis for the new waste packaging policy. After five years of operation, the results of Eco-Emballages still have some way to go to attain the targets in the time agreed (Eco-Emballages, 1998). In 1997 Eco-Emballages achieved a valorisation level of 1.8 million tonnes of packaging waste. The overall valorisation rate (recycling and incineration) reached 31% (22% for recycling alone), with a target of 75% in 2002.

In Greece implementation problems are radically different. Illegal landfilling is still the implicit 'standard' in municipal waste management in many places. Greek industry, in agreement with the public authorities, has taken an initiative to experiment with a system of separate collection resembling the concept of the French Eco-Emballages. The limited scale of the trial in 1998 does not yet allow any comment on the results obtained.

Implementation as a learning process allowing for a re-appraisal of initial choices

Obstacles met at the implementation stage required each form of national regime to make adjustments and introduce revisions of the instruments in order to improve progress in meeting objectives, sometimes reconsidering and changing the objectives themselves. Thus implementing new regimes for waste management also meant learning and adapting initial concepts. What lessons can be derived from the adaptability of various regimes? Two questions deserve consideration: to what extent do systems refusing central command and control prescriptions and based on decentralized decision-making demonstrate a greater facility in introducing changes in the regime through timely corrections? To what extent do corrections that have been introduced increase or reduce the conceptual gap between national regimes? Except for Greece, which is still in the first stage of the elaboration of its regime, the four other countries analysed decided to adjust the initial design on the basis of feed-back from the first years of operation.

In Germany, because of the excessive cost of recycling plastics, a debate took place concerning the quantitative targets that should be imposed on this type of material. Following many negotiations, it was decided to lower the quota for plastic: 60% has to be recovered in total and at least 36% has to be recycled. Such a distinction is new and brings it closer to the French approach, since it gives more room for manoeuvre to a plurality of techniques. Measures have also been taken to prevent the DSD becoming too expensive. Operators are no longer paid on the basis of the input 'waste' they receive but on the output of sorting, and subsidies are progressively increasing on the basis of cost–efficiency ratios. This should reduce the cost of the German regime. Measures have also been taken to limit the export of secondary materials, which had been strongly criticized by other countries and by NGOs. The DKR, responsible for the recycling of plastics, intends to stabilize exports at 10% of the flow and to stop financing plastics recycling in 2003, once valorisation technologies are stabilised. A clear attempt to address the most important sources of problems can therefore be observed in Germany. More attention paid to the impact of the rules of the game on cost–effectiveness and a looser interpretation of the national hierarchy of technologies are two features of this adaptation.

In France, with the renewal of the Eco-Emballages agreement in June 1996, an initial modification was made to the regime; methods of calculating the subsidies to local authorities were revised to favour recycling versus incineration and introduce stronger incentives to encourage the collection and sorting of plastics. Following this revision, an official policy directive from the Ministry of the Environment, on 28 April 1998, modified the goals of the French regime. The directive insisted on a higher place allocated to recycling (including composting), with a long-term projected goal of at least 50% of all household waste. It also clarified the notion of ultimate waste which, until now, was seen by many actors as merely the output of incineration. It specified that they were 'the non-recoverable fraction of waste and not only the output of incineration'. This precision is critical since it encouraged local authorities to move away from incineration and established that, in some clearly defined circumstances, landfilling could be an acceptable solution for waste rejected from sorting or composting activities. Finally, the directive tried to give a more pragmatic interpretation of the proximity principle. Mainly because of the lack of flexibility of incineration capacities, and the required economies of scale, the geographical limits of départements, imposed by the concept of départemental plans of municipal waste management, were often deemed too small an area. Thus the directive called for greater co-operation between adjacent départements on the use of incineration plants for the best environmental and economical benefits. More recently a new round of discussions began on the contributions paid to Eco-Emballages by the industry and on subsidies awarded by Eco-Emballages to local authorities for the separate collection and sorting of waste. New ideas under discussion are an increase in the overall level of support given by Eco-Emballages and differentiated contributions paid according to materials, to achieve an overall financial balance for each material, and a generalization of the progressive mechanism of subsidy that increases the unit rate of support according to the rate of recovery obtained by local authorities. All this will give a new impulse to the national policy of household waste management in the years to come and bring the French regime closer to its German and Dutch counterparts.

In Italy, the regime was fundamentally modified in 1997. The responsibility for the achievements of the objectives was assigned to a new private compulsory consortium, CONAI. This new organization designated two new categories for the packaging producers already involved: the bottlers and the distributors. Its mission was to overcome the main difficulties experienced by the previous Obligatory Syndicates: the lack of involvement of packaging users; this category seemed to be very sensitive to environmental pressure. Since October 1998, packaging producers and bottlers have had to pay a fixed entry fee and a contribution depending on the packaging volumes they introduce onto the market. The revenues from this fee are used to finance collection and sorting operations of packaging waste, displaying some similarities with the DSD and Eco-Emballages. However, the exact role of CONAI and the specific nature of the responsibilities of packaging producers and bottlers are as yet ill-defined and are subject to many criticisms from the Ministry of the Environment. There is at present no guarantee that the management system will be appropriate to the legislation; co-operation between industry and the municipalities is, in addition, still difficult to organize, mainly in the south of the country.

In The Netherlands, it has already been stated that the packaging covenant has been split into many co-covenants in order to better identify the responsibilities for the various materials. Besides this, the proximity principle has been progressively abandoned, to enable demand and supply of waste management to tie in better on a national level. The results are globally satisfying, but the quantitative objectives concern the year 2000, so it is too early to fully assess the overall capacity of the regime to reach the fixed goals.

Whatever the initial features of national regimes, actual experience obliged each of them to pay more attention to the economic dimension of household waste management: incentives embedded with the rules of the game; sensitivity to economies of scale; cost levels imposed by various targets. Those countries starting with a rigid, prescriptive and hierarchical approach have introduced more flexibility into their system to ensure an efficient matching of waste flows to treatment facilities. Those countries which started with a loose approach, confident that local decision making would achieve efficiency, came to establish a more solid framework of targets, rules, organizations and incentives. Hence, initial divergence among countries did not increase with experience but, on the contrary, to some extent decreased.

Key variables in the description of regimes of municipal waste management

The comparative approach of national regimes for household waste management can be usefully employed to extract some bottom-up views on what the key variables are in defining the characteristics of policy regimes. Our own approach led us to select four main variables that determine the essential features of a regime.

The basic conventions supporting the objectives assigned to a new regime

A range of technological fixes exist and, significantly, include, on the one hand, landfilling as an option for the future, not only a heritage from the past and, on the other hand, prevention at source. Three main 'myths' can currently be discerned in Europe, all competing with each other: a 'myth of mastery and containment', of pollutant flows in landfills somehow having no environmental impact; a 'myth of perpetual material cycles' (recycling), able to support economic growth with little external material input and output; a 'myth of a dematerialized post-modern society' (prevention at source).

We call them 'myths' because they do, in reality, inspire certain strategic choices although, at the same time, each one taken in isolation cannot represent a credible option for the future. For instance, if landfilling is adopted as the unique solution, there will unavoidably be environmental spill-over and pollutant leakage quite apart from local problems of social acceptability due to traffic nuisance and site-specific external effects suffered by local populations. Similarly, economic growth can only be sustainable in the long term by mobilising sufficient energy and material flows, though the productivity of these factors is expected to increase. Looking forward to the next decades, we cannot reasonably imagine that a post-modern 'services society' will really be one with a very low material and energy profile! To be frank, the main indicators of the coming 'services society' are the general dissemination of computers with all their hardware and the sharp increase of energy-intensive air traffic; demonstrating that a services society is not necessarily a society possessing a low material and energy profile. Finally, recent national experiences in materials recycling shows how important in practice entropy laws are, expressed in terms of costs: the separate collection and sorting of waste and a concentration of top quality homogenous flows of materials are constraining and costly operations. Good markets for scrap materials cannot be provided beyond reasonable limits.

None of these myths provide a real picture of the society of the future. They just represent attractors in changing technological and institutional equilibria, before their limits are better perceived. Their collective appeal stems from their plurality, promising that there can with certainty be an overall solution; each technological myth provides an escape from difficulties encountered by the two others and all operate in a closed circle.

Policy and technological choices

This variable concerns the way in which governments view their own responsibility and competence in choosing the technological means of achieving policy goals. One view ascribes moral and political attributes to technologies and tends to typify them as good or evil, or somewhere in-between; as if technologies have such attributes for their own sake and not as a result of an assessment of their appropriateness to a specific practical context. A contextual feature is translated in an ontological one. Using this approach, it is possible to define a national hierarchy of technical options that looks like a moral classification of sins. This moral approach to technology has much to commend it in terms of simplicity and political appeal: clear political messages can be given and each local manager automatically knows how to decide which technology to choose.

To this central interventionist and hierarchical approach can be opposed the lack of any general national hierarchy. In this scenario, technological choices are seen as an exercise in instrumental rationality based on cost–benefit calculations, whose results will depend on individual circumstances. No specific measures are accorded intrinsic superiority. The political message is more uncertain and less appealing since it places the responsibility of deciding what to do on the shoulders of local managers, asking them to look for information and perform their own calculations. The French experience has shown that local authorities are not always well equipped in human resources to do this and often prefer to employ standard and uniform solutions, easier to justify than ad hoc, innovative ones.

The patterning of relationships between public authorities and industry in designing a new regime

Two extreme and opposite positions can be discerned: in the first, the design of a new regime is delegated to the regulated agents (companies and local authorities), provided that the government determine the general goals and targets that express of its policy. The hope is that this procedure will prove particularly viable and acceptable among decentralized agents. This responds to a vision of relationships between the government and society based on basic co-operation for the common good, instead of a more conflict-oriented vision in which political action is seen as a struggle and takes the form of an imposition of objectives and rules on society, in opposition to private and vested interests supposed to be hostile by nature to the general interest. The latter view may articulate a prophetic or ethical approach to politics and policy making for which the common good, once it has been identified by inspired politicians and their experts, has to be imposed in a disciplined way on the whole society. In practical terms, this can lead to an imposition of a new regime by the government without a dialogue with vested interests and stakeholders. Each approach has its own drawbacks: a risk of capture of the regulator by the regulated and incentive failures with the first approach; a risk of implementation failure due to the tacit or explicit opposition of the regulatees in the second. The issue must be tackled in terms of the nature of representation and choice of procedures, in order to find a satisfactory equilibrium.

The division of practical and financial responsibilities between actors

Who is to take initiatives and bear responsibilities? Who has the technical know-how to manage material flows? Who should pay at the various stages in the decision-making process in which waste materials are generated? These questions always have two dimensions: a distributive dimension and an incentive dimension. Exempting some actors from any payments and responsibility for distributive reasons (for instance to maintain or improve the competitiveness of some industrial activities) may negatively affect incentives and make the whole system incapable of reaching targets or generate excessive costs. According to the economic interpretation of the polluter pays principle, decision makers involved in environmental problems that require public action should cover the costs involved to reach collective objectives. The purpose of invoking this principle is to encourage decision makers to take account of appropriate economic intelligence when they make their decisions (Turner and Pearce, 1994).

Taken together, these four variables have a good deal to say about the degree of orientation of regimes towards cost-effectiveness. For instance, a policy aiming at the prevention of waste production at source and ambitious objectives for valorisation require important changes in the behaviour of actors. This may be obtained through ambitious targets and extensive financing that acquire an incentive dimension even without the explicit intention of so doing. Meanwhile, an in-depth dialogue favours the acceptance of a policy but also means a regime may neutralize the incentive structure for competing industrial activities, as in the case of the French policy (Defeuilley and Godard, 1997; Godard, 1999).

In France, industrialists played a leading role in the formulation and the implementation of the packaging waste regime. Eco-Emballages was the result of a quasi-delegation of the design of the packaging waste regime to industry. This rather unusual approach in France was intended to encourage all industrial companies to become part of a common endeavour to promote waste valorisation in spite of conflicting economic interest as regards markets. Bridging the gap between bottlers and material producers became a key issue. This consensus-driven process had important initial drawbacks. First, the initial rate of contributions paid by bottlers was not differentiated according to materials but only by volume, to avoid disturbing the pre-existing sharing of markets between materials. In this way bottlers could also ensure they would continue to be allowed to use the whole range of packaging materials without any restrictions. Second, industry came up with scheme that would minimize the financial burden they would bear, by promoting the concept of 'shared responsibility' with local authorities and 'subsidies for supplementary cost' for selective collection and sorting. In several regards, this amounted to a transfer of uncertainties and technological and financial risks to municipalities. Third, industry attempted to gain maximum control over their involvement in collection. As a result the green dot put on household packaging lost almost all incentive value.

The persistent difficulties encountered in Italy in fully involving municipalities are also related to weaknesses in planning, which are themselves related to the policy-making process.

These variables also explain the variation in direct running costs between countries; for example between the German and the French regimes (Figure 1). The attribution of responsibility for waste management and the choices made regarding techniques are the main variables explaining the huge discrepancy in financial results between the two organizations, the DSD and Eco-Emballages. A regime combining ambitious objectives, technical interventionism and the division of the waste flow into separate parts can only lead to a strong increase of the direct costs born by private and public actors.

Policy principles

In addition to identifying an underlying conventional wisdom, which we termed 'myths' in the previous section, several principles play a central organizational role in the regimes adopted by the five European countries. In some cases these principles were common to all countries, but were not interpreted in the same way. Some principles became part of European law through international treaties (Maastricht, Amsterdam), but they did not possess the same characteristics in each country. Interestingly, national interpretations of these principles have also been adjusted and evolution has depended on the history of implementation in each country.

The following principles are relevant to waste management: 'valorisation',

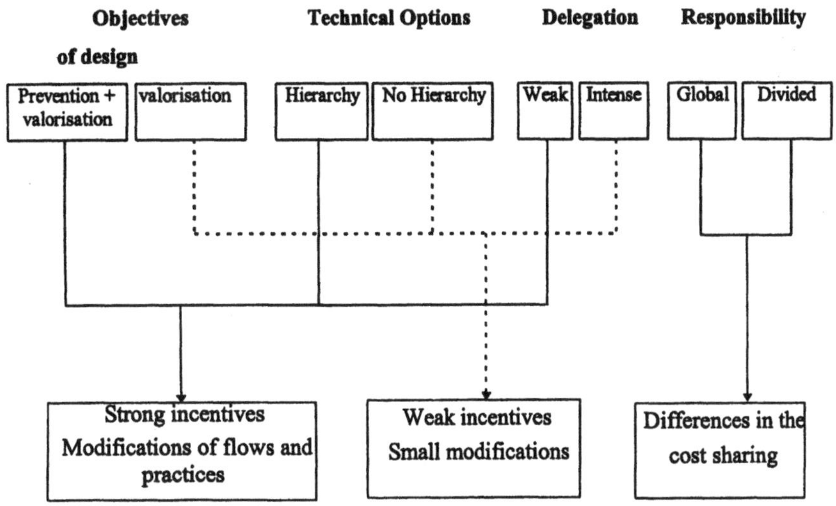

Figure 1. Key variables for describing national regimes

'prevention', 'proximity', 'self-sufficiency' and the 'polluter pays'. Examining the articulation of these principles in each case and the differences in their interpretation may help us understand compatibility problems between national regimes.

The valorisation principle is common to all countries. However, it is not in practice applied in the same way everywhere. Generally speaking, valorisation includes material recycling, energy recovery and composting, each area is not accorded the same status in all countries. Material recycling is undeniably a strong common thread in any regime. Composting, with a similar time-lag and delay in acquiring a political orientation, has achieved the same importance in The Netherlands, Germany and France. Energy recovery is in a different position. In France, Italy and Greece, even if there are some minor legislative signs showing that energy recovery is not treated in exactly the same way as recycling in an implicit hierarchy (see the European Directive 62/94/CE), this valorisation technique is institutionally favoured as much as others.[9] In Germany, the official position is different. In 1992 incineration with energy recovery was forbidden for packaging waste. Even if the German position has been modified, recycling is still the priority in the hierarchy of techniques. Nevertheless, 'chemical recycling', a typical German notion for designating the use of plastics in some industrial processes (steel production), looks very much like incineration with energy recovery. Terminological innovation had to confront a discrepancy between policy principles as defined in law, to which political leaders were committed, and economic and technological realities (how to find an adequate market for scrap plastics?). This is directly related to the German interpretation of the valorisation principle, which identifies incineration, or any technique resembling incineration, as an inferior technique that should be avoided for packaging waste. Such a hierarchy is also present in The Netherlands, but has been recently modified to permit a more efficient use of existing incinerators.

The question of whether there is a national hierarchy of technological solutions invests a different value in the valorisation principle. Where such a hierarchy exists, it means that the principle of prevention is placed above other solutions (Germany, The Netherlands). In this case, the valorisation principle receives a lower ranking in the hierarchy. But it should be stressed that reference to prevention is almost non existent in France, in Italy or in Greece.

The proximity principle and the self-sufficiency principle are closely related. They are partly an answer to the NIMBY syndrome, common to the five countries, which makes it difficult to accept waste coming from other regions and from abroad. They also constitute an attempt to ensure that a sufficient supply of local waste reaches local treatment infrastructures, even if the latter provides services more costly than other competing solutions over a larger area. Economically speaking, what is at stake is organizing the 'capture' of local waste producers so as to ensure the financial viability of investments by local authorities or by appointed private companies. Regulators initially set much store by these principles. In France for example, both the proximity and the self-sufficiency principles were applied at the départemental level, while it was applied at the provincial level in The Netherlands. A similar territorial distribution is observable elsewhere. However, this geographical distribution has proved to be a source of economic inefficiency. The Netherlands have been the first to abandon it, whilst in France the national regulator tends to favours co-operation in waste management between départements. The geographical scale of the operation was enlarged to allow the actors to achieve a closer relationship between waste flows, existing infrastructures and required investment.

The polluter pays principle is also present in each country, with very different interpretations and implementations. Whilst Germany tends, through the DSD, to invoke the producer's responsibility to its full extent with a total cost system, most other countries have institutionalised a shared financial responsibility between industry and local authorities. In this scenario, the public finances approach takes the lead at the expense of a policy of integration of environmental objectives in the decision-making of all relevant actors of the waste system. Variations in the implementation of the polluter pays principle may have a significant impact on industry. In the case of a partial application, industrial companies may derive undue competitive advantages in comparison with competitors having to pay the full cost of waste treatment. There is a direct link between the prevention principle and the polluter pays principle. For example, when bottlers carry a high financial burden, they receive an economic incentive to develop prevention at source. Choosing eco-packages would reduce the financial burden they are faced with. Hence the complete implementation of the polluter pays principle may be seen as a critical step towards the implementation of a prevention policy.

From this overview of conventional principles in the five countries, it is possible to arrive at some final observations. One common point of departure for national regimes is the issue of prevention. Only a few European countries have selected prevention as their prime objective, whilst others prioritize material recycling or valorisation, and still others place controls on landfilling. Does this mean that all countries are going to follow the same road, at their own pace, with some countries acting as standard bearers and others following in the rear guard? Attractive though it might be, such an interpretation, reminiscent of Rostow's steps of economic growth, would be rather superficial, since it ignores the influence of the three alternate 'myths' that underlie strategies of waste management: none of these myths can be identified as the appropriate one for the future while others would be attached to the past. Such a situation should form the basis for a new approach to European harmonisation of national regimes of waste management. Will the prevention of waste generation at source become a European goal or, on the basis of the subsidiarity principle, should it remain a specific objective of individual countries? In the latter case, how can we avoid prevention producing a competitive disadvantage for countries making such a choice? Clearly the harmonization issue is another story, as was shown by the process of negotiation of the EU directive on packaging waste (Buclet and Godard, 1997).

Notes

1. The authors wish to thank Christophe Defeuilley for his precious help.
2. NIMBY: not in my backyard.
3. Valorisation is a generic term including recycling, composting, incineration with energy recovery and any other form of energy recovery (methanisation, co-generation, etc.). In April 1998, a preliminary objective of 50% recycling for household waste was announced by the French Ministry of the Environment. No deadline was set.
4. Ultimate waste includes any waste, whether resulting or not from a form of treatment, the pollutant charge of which cannot be reduced by additional treatments and cannot be valorised, under prevailing technical and economical conditions.
5. For plastics, recycling quotas have been lowered. 60% has to be recovered and at least 36% has to be recycled (material recycling). Chemical recycling is therefore capped at a maximum of 24%.
6. Greece, Ireland and Portugal benefit from specific and less stringent measures.
7. Objectives are fixed at a more general level, extending to the whole waste flow. The management of municipal waste forms an integral part of these objectives.
8. This differentiation is subject to the sorting rate achieved. The financial support for material sorting indeed increases with it.
9. This position has, however, changed in France in 1998.

References

Bertolini G., 1994, Economies sur l'incinération d'ordures ménagères résultant d'une réduction à la source ou de collectes sélectives (notamment de 'fractions sèches' et en particulier d'emballages), TSM, 89.

Buclet N. and O. Godard, 1997, L'emergence négociée d'une norme économique commune au sein des institutions européennes: la directive communautaire sur les emballages et les déchets d'emballages, Revue Politiques et Management Public, 15, 73–95.

Coopers and Lybrand, 1995, Impact économique des mesures prises en faveur de la valorisation des déchets d'emballages, Rapport pour le Ministère de l'Industrie, Paris.

Defeuilley C. and O. Godard, 1997, The economics of packaging waste recycling in France: institutional framework and technological adoption, International Journal of Environment and Pollution, 7, 538–546.

Defeuilley C., S. Lupton and Y. Serret, 1997, The French household waste regime, Report for DG XII, European Commission, Paris: CIRED.

Defeuilley C. and P. Quirion, 1995, Les déchets d'emballages ménagers: une analyse économique des politiques frañaise et allemande, Economie et Statistique, 290.

Denby-Wilkes, V., 1996, Politique nationale: sa traduction en termes d'équipements et d'investissements. Conférence La Gestion des déchets municipaux: comment mieux gérer au moindre coût, Paris, 26 June.

Dente B., ed., 1995, Environmental Policy in Search of New Instruments, Dordrecht, Kluwer Academic Press.

Dente B., P. Fareri and J. Ligteringen, eds, 1998, The Waste and the Backyard. The Creation of Waste Facilities: Success Stories in Six European Countries, Dordrecht, Kluwer Academic Press.

Department of the Environment, 1992, Economic Instruments and Recovery of Resources from Waste, HMSO, London.

Duales System Deutschland, 1995, Packaging Recycling Worldwide, Cologne.

Eco-Emballages, 1998, Yearly Report, Levallois-Perret.

Erkman, S., 1998, Vers une écologie industrielle, Paris, Editions Charles Léopold Mayer.

Godard O., 1999, Dialogue and economic efficiency: two antagonistic goals for environmental policy-making? Lessons from the French packaging-waste management system, in Skou Andersen, M. and R. Sprenger, eds, Designing, Introducing and Implementing Market-based Instruments in Environmental Management: Institutional Dimensions, Cheltenham, Edward Elgar (forthcoming).

Leppänen, A., 1994, The experience of the Scandinavian countries, 3rd European Congress Packaging & Environment, Brussels, 21–22 November.

Neumann F., 1997, Municipal waste management in The Netherlands, Report for DG XII, European Commission, Rotterdam: ERASMUS.

Royal Commission on Environmental Pollution, 1993, Seventeenth Report: Incineration of Waste, HMSO, London.

Turner R. and D. Pearce, 1994, The role of economic instruments in solid waste management policy, in Opschoor J. and R. Turner, eds., Economic Incentives and Environmental Policies: Principle and Practice, Dordrecht, Kluwer Academic Press, 251–273.

von Weizsäcker E., A.B. Lovins and L.H. Lovins, 1997, Factor Four. Doubling Wealth Halving Resource Use, London, Earthscan Pub.

Whiston T. and M. Glachant, 1996, Voluntary agreements between industry and government. The case of recycling regulations, in Levêque F., ed., Environmental Policy in Europe. Industry, Competition and the Policy Process, Cheltenham, Edward Elgar.

List of contributors

Andrea Andreadakis, Department of Water Resources, Athens, Greece

Nicolas Buclet, Centre International de Recherche sur l'Environnement et le Développement, Nogent sur Marne, France

Lilo Fischer, Institut für Ökologische Wirtschaftsforschung, Berlin, Germany

Olivier Godard, Laboratoire d'Econométrie, Ecole Polytechnique, Paris, France

Win Hafkamp, Erasmus Center for Environmental Studies, Rotterdam, The Netherlands

Antonio Massarutto, Instituto di Economia delle Fonti di Energia, Milan, Italy

Index

Environment & Management

1. O. Kuik and H. Verbruggen (eds.): *In Search of Indicators of Sustainable Development.* 1991 ISBN 0-7923-1249-X
2. R. Janssen: *Multiobjective Decision Support for Environmental Management.* 1992 ISBN 0-7923-1908-7
3. R. Janssen and M. van Herwijnen: *DEFINITE. A System to Support DEcisions on a FINITE Set of Alternatives.* Software with User Manual. 1994 ISBN 0-7923-2696-2
4. D. James: *The Application of Economic Techniques in Environmental Impact Assessment.* 1994 ISBN 0-7923-2721-7
5. P. Glasbergen (ed.): *Managing Environmental Disputes: Network Management as an Alternative.* 1995 ISBN 0-7923-3034-X
6. L. Reijnders (ed.): *Environmentally Improved Production Processes and Products: An Introduction.* 1996 ISBN 0-7923-3786-7
7. E. Beinat: *Value Functions for Environmental Management.* 1997 ISBN 0-7923-4662-9
8. B. Dente, P. Fareri and J. Ligteringen (eds.): *The Waste and the Backyard.* The Creation of Waste Facilities: Success Stories in Six European Countries. 1998 ISBN 0-7923-5028-6
9. E. Beinat and P. Nijkamp (eds.): *Multicriteria Analysis for Land-Use Management.* 1998 ISBN 0-7923-5198-5
10. N. Buclet and O. Godard (eds.): *Municipal Waste Management in Europe.* A Comparative Study in Building Regimes. 1999 ISBN 0-7923-5885-6

KLUWER ACADEMIC PUBLISHERS – DORDRECHT / BOSTON / LONDON

MIX
Papier aus verantwortungsvollen Quellen
Paper from responsible sources
FSC® C105338

If you have any concerns about our products,
you can contact us on
ProductSafety@springernature.com

In case Publisher is established outside the EU,
the EU authorized representative is:
Springer Nature Customer Service Center GmbH
Europaplatz 3, 69115 Heidelberg, Germany

Printed by Libri Plureos GmbH
in Hamburg, Germany